Modern Birkhäuser Classics

Many of the original research and survey monographs, as well as textbooks, in pure and applied mathematics published by Birkhäuser in recent decades have been groundbreaking and have come to be regarded as foundational to the subject. Through the MBC Series, a select number of these modern classics, entirely uncorrected, are being re-released in paperback (and as eBooks) to ensure that these treasures remain accessible to new generations of students, scholars, and researchers.

T0192094

Introduction to Stochastic Integration

K.L. Chung
R.J. Williams

Second Edition

Reprint of the 1990 Edition

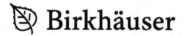 Birkhäuser

K.L. Chung (deceased)
Department of Mathematics
Stanford University
Stanford, CA, USA

R.J. Williams
Department of Mathematics
University of California at San Diego
La Jolla, CA, USA

Originally published in the series *Probability and Its Applications*

ISSN 2197-1803   ISSN 2197-1811 (electronic)
ISBN 978-1-4614-9586-4   ISBN 978-1-4614-9587-1 (eBook)
DOI 10.1007/978-1-4614-9587-1
Springer New York Heidelberg Dordrecht London

Library of Congress Control Number: 2013953549

Printed on acid-free paper

Springer is part of Springer Science+Business Media (www.birkhauser-science.com)

K.L. Chung R.J. Williams

Introduction to Stochastic Integration

Second Edition

Birkhäuser
Boston • Basel • Berlin

K.L. Chung
Department of Mathematics
Stanford University
Stanford, California 94305, USA

R.J. Williams
Department of Mathematics
University of California at San Diego
La Jolla, California 92093, USA

Cover Image. A Brownian motion B starts at a point x inside the domain D and first leaves D at B_τ. Under conditions explained in Section 6.4, a solution to the Schrödinger equation $\frac{1}{2} \Delta \psi + q\psi = 0$ in D that approaches f on the boundary of D can be represented by

$$\psi(x) = E^x \left[\exp \left(\int_0^\tau q(B_t)\, dt \right) f(B_\tau) \right].$$

The Brownian path used in this illustration is from The Fractal Geometry of Nature © 1982 by Benoit B. Mandelbrot and is used with his kind permission.

Library of Congress Cataloging-in-Publication Data
Chung, Kai Lai, 1917–
   Introduction to stochastic integration / K.L. Chung, R.J.
Williams. — 2nd ed.
   p. cm. — (Probability and its applications)
   Includes bibliographical references (p.   ) and index.
   ISBN 0-8176-3386-3 (U.S. : acid-free paper). — ISBN 3-7643-3386-3
(Switz. — acid-free paper)
   1. Integrals, Stochastic. 2. Martingales (Mathematics)
I. Williams, R. J. (Ruth J.), 1955–   . II. Title. III. Series.
QA274.22.C48   1990
519.2—dc20   90-1020
                                                       CIP

Printed on acid-free paper.

ISBN 0-8176-3386-3
ISBN 3-7643-3386-3

Camera-ready copy provided by the authors using T_EX.
Printed and bound by Edwards Brothers, Inc., Ann Arbor, Michigan.
Printed in the U.S.A.

9 8 7 6 5 4 3

PREFACE

This is a substantial expansion of the first edition. The last chapter on stochastic differential equations is entirely new, as is the longish section §9.4 on the Cameron-Martin-Girsanov formula. Illustrative examples in Chapter 10 include the warhorses attached to the names of L. S. Ornstein, Uhlenbeck and Bessel, but also a novelty named after Black and Scholes. The Feynman-Kac-Schrödinger development (§6.4) and the material on reflected Brownian motions (§8.5) have been updated. Needless to say, there are scattered over the text minor improvements and corrections to the first edition. A Russian translation of the latter, without changes, appeared in 1987.

Stochastic integration has grown in both theoretical and applicable importance in the last decade, to the extent that this new tool is now sometimes employed without heed to its rigorous requirements. This is no more surprising than the way mathematical analysis was used historically. We hope this modest introduction to the theory and application of this new field may serve as a text at the beginning graduate level, much as certain standard texts in analysis do for the deterministic counterpart. No monograph is worthy of the name of a true textbook without exercises. We have compiled a collection of these, culled from our experiences in teaching such a course at Stanford University and the University of California at San Diego, respectively. We should like to hear from readers who can supply more and better exercises.

A word about the exposition. We have consistently chosen clarity over brevity. As one of the authors suggested elsewhere, readers who insist on concision are free to skip every other line or so. But be warned that most errors in mathematics are concealed under the surreptitious cover of terseness, whereas a fuller exposure leaves less to pitfalls. A good example of our preference is afforded by the demonstration in §10.3 of the Markov property for the family of solutions of a stochastic differential equation, which is often glossed over in texts and *ergo* gloated over by innocent readers. Actually, the point at issue there is subtle enough to merit the inculcation.

For the new material, the following acknowledgements are in order. Michael Sharpe provided helpful comments on several points in Chapter 2. Giorgio Letta inspired an extension of predictable integrability in Chapter 3. Martin Barlow supplied two examples in Chapter 3. Daniel Revuz and Marc Yor permitted the references to their forthcoming book. Darrell Duffie gave lectures on the Black-Scholes model in Chung's class during 1986 which led to its inclusion in §10.5. We wish also to thank those colleagues and students who contributed comments and corrections to the first edition. Lisa Taylor helped with the proof reading of this edition. Kathleen Flynn typed the manuscipt of the first edition in TEX, whilst artists at Stanford Word Graphics, especially Walter Terluin, added final touches to the figures. We are appreciative of the interest and cooperation of the staff at Birkhäuser Boston. Indeed, the viability of this new edition was only an optional, not a predictable event when we prepared its precursor in 1983.

March 1990   K. L. Chung
                                                R. J. Williams

PREFACE
TO THE FIRST EDITION

The contents of this monograph approximate the lectures I gave in a graduate course at Stanford University in the first half of 1981. But the material has been thoroughly reorganized and rewritten. The purpose is to present a modern version of the theory of stochastic integration, comprising but going beyond the classical theory, yet stopping short of the latest discontinuous (and to some distracting) ramifications. Roundly speaking, integration with respect to a local martingale with continuous paths is the primary object of study here. We have decided to include some results requiring only right continuity of paths, in order to illustrate the general methodology. But it is possible for the reader to skip these extensions without feeling lost in a wilderness of generalities. Basic probability theory inclusive of martingales is reviewed in Chapter 1. A suitably prepared reader should begin with Chapter 2 and consult Chapter 1 only when needed. Occasionally theorems are stated without proof but the treatment is aimed at self-containment modulo the inevitable prerequisites. With considerable regret I have decided to omit a discussion of stochastic differential equations. Instead, some other applications of the stochastic calculus are given; in particular Brownian local time is treated in detail to fill an unapparent gap in the literature. The applications to storage theory discussed in Section 8.4 are based on lectures given by J. Michael Harrison in my class. The material in Section 8.5 is Ruth Williams's work, which has now culminated in her dissertation [77].

At the start of my original lectures, I made use of Métivier's lecture notes [59] for their ready access. Later on I also made use of unpublished notes on continuous stochastic integrals by Michael J. Sharpe, and on local time by John B. Walsh. To these authors we wish to record our indebtedness. Some oversights in the references have been painstakingly corrected here. We hope any oversight committed in this book will receive similar treatment.

A methodical style, due mainly to Ruth Williams, is evident here. It is not always easy to strike a balance between utter precision and relative readability, and the final text represents a compromise of sorts. As a good author once told me, one cannot really hope to achieve consistency in writing a mathematical book—even a small book like this one.

December 1982   K. L. Chung

TABLE OF CONTENTS

ABBREVIATIONS AND SYMBOLS

PRELIMINARIES

1.1 Notations and Conventions

For each interval I in $\mathbb{R} = (-\infty, \infty)$ let $\mathcal{B}(I)$ denote the σ-field of Borel subsets of I. For each $t \in \mathbb{R}_+ = [0, \infty)$, let $\mathcal{B}_t$ denote $\mathcal{B}([0,t])$ and let $\mathcal{B}$ denote $\mathcal{B}(\mathbb{R}_+) = \bigvee_{t \in \mathbb{R}_+} \mathcal{B}_t$ — the smallest σ-field containing $\mathcal{B}_t$ for all t in $\mathbb{R}_+$. Let $\overline{\mathbb{R}}_+ = [0, \infty]$ and $\overline{\mathcal{B}}$ denote the Borel σ-field of $\overline{\mathbb{R}}_+$ generated by $\mathcal{B}$ and the singleton $\{\infty\}$. Let λ denote the Lebesgue measure on $\mathbb{R}$.

Whenever t appears without qualification it denotes a generic element of $\mathbb{R}_+$. The collection $\{x_t, t \in \mathbb{R}_+\}$ is frequently denoted by $\{x_t\}$. The parameter t is sometimes referred to as time.

Let $\mathbb{N}$ denote the set of natural numbers, $\mathbb{N}_0$ denote $\mathbb{N} \cup \{0\}$, and $\mathbb{N}_\infty$ denote $\mathbb{N} \cup \{\infty\}$. Whenever n, k, or m, appears without qualification, it denotes a generic element of $\mathbb{N}$. A sequence $\{x_n, n \in \mathbb{N}\}$ is frequently denoted by $\{x_n\}$. We write $x_n \to x$ when $\{x_n\}$ converges to x. A sequence of real numbers $\{x_n\}$ is said to be increasing (decreasing) if $x_n \le x_{n+1}$ ($x_n \ge x_{n+1}$) for all n. The notation $x_n \uparrow x$ ($x_n \downarrow x$) means $\{x_n\}$ is increasing (decreasing) with limit x.

For each $d \in I\!N$, the components of $x \in I\!R^d$ are denoted by x_i, or sometimes by x^i, $1 \leq i \leq d$, and the Euclidean norm of x is denoted by $|x| = \left(\sum_{i=1}^{d} (x_i)^2 \right)^{\frac{1}{2}}$.

The symbol 1_A denotes the indicator function of a set A, i.e., $1_A(x) = 1$ if $x \in A$ and $= 0$ if $x \notin A$. The symbol $\emptyset$ denotes the empty set.

For each n, $C^n(I\!R)$ or simply C^n denotes the set of all real-valued continuous functions defined on $I\!R$ for which the first n derivatives exist and are continuous. We use $C(I\!R)$ to denote the set of real-valued continuous functions on $I\!R$ and $C^\infty(I\!R)$ or C^∞ to denote $\bigcap_{n \in I\!N} C^n$, the set of infinitely differentiable real-valued functions on $I\!R$.

We use the words "positive", "negative", "increasing", and "decreasing", in the loose sense. For example, "x is positive" means "$x \geq 0$"; the qualifier "strictly" is added when "$x > 0$" is meant. The infimum of an empty set of real numbers is defined to be ∞. A sum over an empty index set is defined to be zero.

1.2 Measurability, L^p Spaces and Monotone Class Theorems

Suppose (S, Σ) is a measurable space, consisting of a non-empty set S and a σ-field Σ of subsets of S. A function $X : S \to I\!R^d$ is called Σ-measurable if $X^{-1}(A) \in \Sigma$ for all Borel sets A in $I\!R^d$, where X^{-1} denotes the inverse image. A similar definition holds for a function $X : S \to \overline{I\!R} = [-\infty, \infty]$. We use "$X \in \Sigma$" to mean "$X$ is Σ-measurable" and "$X \in b\Sigma$" to mean "X is bounded and Σ-measurable".

If Γ is a sub-family of Σ, a function $X : S \to I\!R^d$ is called Γ-*simple* if $X = \sum_{k=1}^{n} c_k 1_{\Lambda_k}$ for some constants c_k in $I\!R^d$, sets $\Lambda_k \in \Gamma$, and $n \in I\!N$. Such a function is Σ-measurable. Conversely, any Σ-measurable function is a pointwise limit of a sequence of Σ-simple functions. For example, a Σ-measurable function $X : S \to I\!R$ is the pointwise limit of the sequence

$\{X^n\}$ of Σ-simple functions defined by

$$X^n = \sum_{k=0}^{n2^n} \frac{k}{2^n} 1_{\{k2^{-n} \leq X < (k+1)2^{-n}\}}$$

$$+ \sum_{k=-1}^{-n2^n} \frac{(k+1)}{2^n} 1_{\{k2^{-n} \leq X < (k+1)2^{-n}\}}$$

and $|X^n| \uparrow |X|$. In the above we have suppressed the argument of X, as we often do in the text.

Suppose ν is a (positive) measure on (S, Σ). A set in Σ of ν-measure zero is called a ν-null set. For $p \in [1, \infty)$, $L^p(S, \Sigma, \nu)$ denotes the vector space of Σ-measurable functions $X : S \to \mathbb{R}$ for which

$$\|X\|_p \equiv \left(\int_S |X(s)|^p \, \nu\,(ds) \right)^{\frac{1}{p}}$$

is finite. We use "ν-a.e." to denote "ν-almost everywhere". If functions which are equal ν-a.e. are identified, then $L^p(S, \Sigma, \nu)$ is a Banach space with norm $\| \cdot \|_p$. In the case $p = 2$, it is also a Hilbert space with inner product $(\cdot , \cdot)$ given by $(X, Y) = \int_S X(s)Y(s)\nu\,(ds)$ for X and Y in $L^2(S, \Sigma, \nu)$. Whenever we view these spaces in this way, it will be implicit that we are identifying functions which are equal ν-a.e.

Monotone class theorems constitute one of the most useful tools in measure theory. These theorems are used for extending certain properties that are readily verified for a small class of sets or functions to a larger class generated by the smaller one. Versions of the monotone class theorems, one for sets and one for functions, are stated below for later reference. See Blumenthal and Getoor [5, pp. 5–6] for proofs of these results.

Let $\mathcal{A}$ be a collection of subsets of a set Γ such that $\mathcal{A}$ is closed under finite intersections. The σ-field generated by $\mathcal{A}$ is denoted by $\sigma(\mathcal{A})$.

Monotone class theorem for sets. *Let S be a collection of subsets of Γ satisfying the following three properties.*

(i) $\Gamma \in S$.

(ii)   *If $A, B \in S$ and $A \subset B$, then $B \backslash A \in S$.*

(iii)   *If $A_n \in S$ and $A_n \uparrow A$, then $A \in S$.*

Under these assumptions, if $A \subset S$ then $\sigma(A) \subset S$.

In the following, a vector space of real-valued functions on Γ is a set V of functions $f : \Gamma \rightarrow I\!R$ such that

(i) $f + g$, defined by $(f + g)(x) = f(x) + g(x)$ for all $x \in \Gamma$, is in V whenever f and g are in V, and

(ii) αf, defined by $(\alpha f)(x) = \alpha f(x)$ for all $x \in \Gamma$, is in V for all $\alpha I\!R$ and $f \in V$.

Monotone class theorem for functions. *Let V be a vector space of real-valued functions on Γ satisfying the following two properties.*

(i) $1_\Gamma \in V$ *and* $1_A \in V$ *for all* $A \in \mathcal{A}$.

(ii)   *If $\{f_n\}$ is an increasing sequence of non-negative functions in V such that $f \equiv \sup_n f_n$ is finite (bounded), then $f \in V$.*

Under these assumptions, V contains all real-valued (bounded) functions on Γ that are $\sigma(\mathcal{A})$-measurable.

1.3 Functions of Bounded Variation and Stieltjes Integrals

For a real-valued function g on $I\!R_+$, the *variation* of g on $[0, t]$ is given by

$$|g|_t \equiv \sup \left(\sum_{k=0}^{n-1} |g(t_{k+1}) - g(t_k)| \right)$$

where the supremum is over all partitions $0 = t_0 < t_1 < \ldots < t_n = t$ of $[0, t]$. The variation $|g|_t$ is increasing in t. If $|g|_t < \infty$, g is said to be of *bounded variation* on $[0, t]$. If this is true for all t in $\mathbb{R}_+$, g is said to be *locally of bounded variation* on $\mathbb{R}_+$; and if $\sup_{t \in \mathbb{R}_+} |g|_t < \infty$, then g is of *bounded variation* on $\mathbb{R}_+$. A (continuous) function is locally of bounded variation on $\mathbb{R}_+$ if and only if it is the difference of two (continuous) increasing functions (see Royden [69, p. 100]).

For $t \in \mathbb{R}_+$, a right continuous function g which is of bounded variation on $[0, t]$ induces a signed measure μ on $([0, t], \mathcal{B}_t)$:

$$\mu((a, b]) = g(b) - g(a) \text{ for } 0 \leq a < b \leq t \text{ and } \mu(\{0\}) = 0.$$

The right continuity of g is used in verifying that μ is a measure on $([0, t], \mathcal{B}_t)$, and μ is uniquely determined since intervals of the form $(a, b]$ together with $\{0\}$ generate $\mathcal{B}_t$. If g is continuous, then μ has no atoms. If g is right continuous and locally of bounded variation on all of $\mathbb{R}_+$, then the measures μ for different intervals $[0, t]$ are consistent with one another, but they do not in general extend to a measure on $\mathbb{R}_+$. However, if g is of bounded variation on $\mathbb{R}_+$, then the measures extend to a finite signed measure on $\mathbb{R}_+$; if g is increasing on $\mathbb{R}_+$, they extend to a positive σ-finite measure on $\mathbb{R}_+$. Returning now to the case of a right continuous g that is of bounded variation on a fixed interval $[0, t]$, we note that the variation $|\mu|$ of μ is the measure associated with the variation $|g|$ of g. If $f \in L^1([0, t], \mathcal{B}_t, |\mu|)$, then the Lebesgue-Stieltjes integral of f with respect to g over $[0, t]$ is defined by

$$\int_{[0,t]} f(s)\, dg(s) \equiv \int_{[0,t]} f\, d\mu$$

$$= \lim_{n \to \infty} \left(\sum_{k=0}^{\infty} \frac{k}{2^n} \mu\left(\left\{ s \in [0, t] : \frac{k}{2^n} \leq f(s) < \frac{k+1}{2^n} \right\}\right) \right.$$

$$\left. + \sum_{k=-1}^{-\infty} \frac{(k+1)}{2^n} \mu\left(\left\{ s \in [0, t] : \frac{k}{2^n} \leq f(s) < \frac{k+1}{2^n} \right\}\right) \right)$$

and $\left| \int_{[0,t]} f(s)\, dg(s) \right| \leq \int_{[0,t]} |f|\, d|\mu|$. If the last integral is finite for all $t \in [0, T]$ and g is continuous, then $\int_{[0,t]} f(s)\, dg(s)$ is a continuous function of $t \in [0, T]$ and we denote it by $\int_0^t f(s)\, dg(s)$. If f is a continuous function

on $[0, t]$, then the Riemann-Stieltjes integral of f with respect to g on $[0, t]$
is well defined and equals the Lebesgue-Stieltjes integral, i.e.,

$$(1.1) \qquad \int_{[0,t]} f(s)\, dg(s) = \lim_{n \to \infty} \sum_{k=1}^{N_n} f(s_k^n) \left(g(t_k^n) - g(t_{k-1}^n) \right),$$

for any sequence of partitions $0 = t_0^n < t_1^n < \ldots < t_{N_n}^n = t$ of $[0, t]$ where
$s_k^n \in [t_{k-1}^n, t_k^n]$ and $\max_{k=1}^{N_n} |t_k^n - t_{k-1}^n| \to 0$ as $n \to \infty$. On the other
hand, if g is continuous as well as being of bounded variation on $[0, t]$, then
$\int_0^t f(s)\, dg(s)$ is also given by (1.1) when f is right continuous on $[0, t)$ with
finite left limits on $(0, t]$, or is left continuous on $(0, t]$ with finite right limits
on $[0, t)$. If f and g are both continuous and of bounded variation on $[0, t]$,
then we have the simple integration by parts formula

$$\int_0^t g(s)df(s) = f(t)g(t) - f(0)g(0) - \int_0^t f(s)dg(s).$$

If f and g are continuous on $[0, t]$, but only g is of bounded variation, then
the right member above is well defined as a Riemann-Stieltjes integral, and
this can be used to show that $\int_0^t g(s)df(s)$ is well defined as a Riemann-
Stieltjes integral, i.e, by the right member of (1.1) with f and g interchanged
there.

1.4 Probability Space, Random Variables, Filtration

Throughout this book, $(\Omega, \mathcal{F}, P)$ denotes a given complete probability
space. This means that $(\Omega, \mathcal{F})$ is a measurable space and P is a probability
measure on $(\Omega, \mathcal{F})$ such that each subset of a P-null set in $\mathcal{F}$ is in $\mathcal{F}$. The
abbreviation "a.s." for "almost surely" means "P-a.e.". The symbol ω
denotes a generic element of Ω. For a function $Y : \Omega \to I\!\!R^d$ (or $\overline{I\!\!R}$) and a
set A in $I\!\!R^d$ (or $\overline{I\!\!R}$), $Y^{-1}(A) = \{\omega : Y(\omega) \in A\}$ is also written as $\{Y \in A\}$.
The symbol ω is also suppressed in similar expressions.

We write L^p for $L^p(\Omega, \mathcal{F}, P)$. For $X \in L^1$, $E(X) \equiv \int_\Omega X\, dP$ denotes
the expectation of X. As an extension of notation, for $\Lambda \in \mathcal{F}$, $E(X ; \Lambda)$
denotes $\int_\Lambda X\, dP$, and when Λ is of the form $\{Y \in A\}$ this is written as
$E(X ; Y \in A)$.

An $\mathcal{F}$-measurable function $X : \Omega \to I\!\!R^d$ is called a *random variable* if $d = 1$ or a *random vector* if $d \geq 2$. The abbreviation "r.v." is used for "random variable". For an arbitrary index set Γ and arbitrary functions X_α from Ω to $I\!\!R^d$ or $\overline{I\!\!R}$, the σ-field $\sigma\{X_\alpha, \alpha \in \Gamma\}$ is the smallest σ-field of subsets of Ω such that X_α is measurable with respect to it for each $\alpha \in \Gamma$. This is called the σ-field generated by the collection $\{X_\alpha, \alpha \in \Gamma\}$. If $\mathcal{G}$ is a sub-σ-field of $\mathcal{F}$, the *augmentation* $\mathcal{G}^\sim$ of $\mathcal{G}$ is the smallest σ-field containing $\mathcal{G}$ and all of the P-null sets in $\mathcal{F}$.

A *filtration* is a family $\{\mathcal{F}_t, t \in I\!\!R_+\}$ of sub-σ-fields of $\mathcal{F}$ such that $\mathcal{F}_s \subset \mathcal{F}_t$ for all $s < t$ in $I\!\!R_+$. If the following two conditions are also satisfied, then $\{\mathcal{F}_t, t \in I\!\!R_+\}$ is called a *standard filtration*:

(i) $\mathcal{F}_t = \mathcal{F}_{t+} \equiv \bigwedge_{s>t} \mathcal{F}_s$ for all t;

(ii) $\mathcal{F}_0$ contains all of the P-null sets in $\mathcal{F}$.

Conditions (i) and (ii), referred to respectively as right continuity and completeness, are not obligatory but simplify many technicalities. We often omit the "$t \in I\!\!R_+$" when referring to a filtration, i.e., we use $\{\mathcal{F}_t\}$.

1.5 Convergence, Conditioning

We review some of the basic notions of probability theory below. It is assumed that readers are familiar with the elementary properties of convergence in L^p, in probability (in pr.), and almost sure (a.s.) convergence, also that of uniform integrability (u.i.) and convergence in distribution (in dist.). See for example Chapter 4 of Chung [11]. For the sake of convenience we state the following proposition.

Proposition 1.1. *Let* $p \in [1, \infty)$ *and* $\{X_n\}$ *be a sequence of random variables in* L^p *which converges in pr. or a.s. to a random variable* X. *Then the following three statements are equivalent.*

(i) $\{X_n\}$ *converges to* X *in* L^p.

(ii) $\{|X_n|^p\}$ *is uniformly integrable.*

(iii) $\lim\limits_{n \to \infty} E(|X_n|^p) = E(|X|^p)$.

Furthermore, if either of the following two conditions is satisfied, then (i)–(iii) hold.

(iv) $\sup_n E(|X_n|^q) < \infty$ *for some* $p < q < \infty$.

(v)   *There is* $Y \in L^p$ *such that* $|X_n| \leq Y$ *for all* n.

The basic properties of conditional expectation (and probability) are also assumed. See for example Chapter 9 of Chung [11]. The conditional expectation of $X \in L^1$ relative to a sub-σ-field $\mathcal{G}$ of $\mathcal{F}$ is denoted by $E(X \mid \mathcal{G})$. This denotes a member of the equivalence class of $\mathcal{G}$-measurable r.v.'s satisfying

$$E\left(E\left(X \mid \mathcal{G}\right); \Lambda\right) = E\left(X; \Lambda\right)$$

for all $\Lambda \in \mathcal{G}$. The following proposition is given for later reference.

Proposition 1.2. *Suppose* $\{X_n\}$ *converges in* L^p *to* $X \in L^p$ *for some* $p \in [1, \infty)$. *Then for any sub-σ-field* $\mathcal{G}$ *of* $\mathcal{F}$, $\{E(X_n \mid \mathcal{G})\}$ *converges in* L^p *to* $E(X \mid \mathcal{G})$.

Proof. This follows by Jensen's inequality for conditional expectations (Theorem 9.1.4 of Chung [11]), which implies

$$E\left(|E(X_n \mid \mathcal{G}) - E(X \mid \mathcal{G})|^p\right) \leq E\left(E\left(|X_n - X|^p \mid \mathcal{G}\right)\right)$$
$$= E(|X_n - X|^p). \quad \blacksquare$$

1.6 Stochastic Processes

A *d-dimensional* (stochastic) *process* X is a function $X : I \times \Omega \to \mathbb{R}^d$ where I is an interval in $\mathbb{R}_+$ and $X(t, \cdot)$ is $\mathcal{F}$-measurable for each $t \in I$. When the dimension is unimportant or is understood, the qualifier "*d*-dimensional" is omitted. A process X is *measurable* if X is $\mathcal{B}(I) \times \mathcal{F}$-measurable. We say that X has *initial value* $x \in \mathbb{R}^d$ if $0 \in I$ and $X(0, \cdot) = x$ a.s. The process X is also denoted by $\{X_t, t \in I\}$, or simply

$\{X_t\}$ when $I = \mathbb{R}_+$. The random variable/vector $X(t, \cdot)$ is also denoted by $X(t)$ or X_t. Given an increasing family $\{\mathcal{F}_t, t \in I\}$ of σ-fields on Ω, the process X is said to be *adapted* to this family if $X_t \in \mathcal{F}_t$ for each $t \in I$. The process X is said to be (right/left) continuous if $t \to X(t, \omega)$ is (right/left) continuous on I for each $\omega \in \Omega$. Of course, right (left) continuity at the right (left) end-point of I is not defined.

Two collections of random variables (or vectors), $X = \{X_\alpha, \alpha \in \Gamma\}$ and $Y = \{Y_\alpha, \alpha \in \Gamma\}$, indexed by the same set, are *versions* of each other if $P(X_\alpha = Y_\alpha) = 1$ for all $\alpha \in \Gamma$. We say X and Y are *indistinguishable* if $P(X_\alpha = Y_\alpha$ for all $\alpha \in \Gamma) = 1$. Two collections which are versions of each other need not be indistinguishable. However, if X and Y are versions of each other and Γ is countable or X and Y are right (or left) continuous processes, then they are indistinguishable. For all practical purposes, indistinguishable processes should be regarded as the same. For instance, if in our definition of continuous processes we required only that *almost* all paths be continuous, such a process would be indistinguishable from one with all paths continuous. Indeed, we shall sometimes prove that on the complement of a P-null set there is a continuous version of a given process. It is trivial to define the version on the null set to make it continuous on all of Ω. Such a version is unique up to indistinguishability. The same is true if we replace continuous by right continuous in the above.

From now on, the term process means a one-dimensional process with $I = \mathbb{R}_+$, unless otherwise stated.

1.7 Optional Times

An $\mathcal{F}$-measurable function $\tau : \Omega \to \overline{\mathbb{R}}_+$ is called an *optional time* with respect to a filtration $\{\mathcal{F}_t\}$ iff $\{\tau \le t\} \in \mathcal{F}_t$ for each $t \in \mathbb{R}_+$. If $\{\mathcal{F}_t\}$ is a standard filtration so that $\mathcal{F}_t = \mathcal{F}_{t+}$, then the condition on τ is equivalent to $\{\tau < t\} \in \mathcal{F}_t$ for each t.

Associated with an optional time τ is a σ-field $\mathcal{F}_\tau$. This consists of all sets A in $\mathcal{F}_\infty \equiv \bigvee_{t \in I\!R_+} \mathcal{F}_t$ satisfying

$$A \cap \{\tau \leq t\} \in \mathcal{F}_t \text{ for all } t \in I\!R_+.$$

We leave it as an exercise to verify that $\mathcal{F}_\tau$ is a σ-field and τ is $\mathcal{F}_\tau$-measurable, and furthermore that if η is an optional time such that $\eta \leq \tau$, then $\mathcal{F}_\eta \subset \mathcal{F}_\tau$.

1.8 Two Canonical Processes

Poisson process. A process $N = \{N_t, t \in I\!R_+\}$ is a *Poisson process* with parameter $\alpha > 0$ if it has the following properties:

(i) $N_0 = 0$,

(ii)   for $0 \leq s < t < \infty$, $N_t - N_s$ is a Poisson r.v. with mean $\alpha(t - s)$, i.e., $N_t - N_s$ takes values in $I\!N_0$ such that

$$P(N_t - N_s = n) = \frac{(\alpha(t - s))^n e^{-\alpha(t-s)}}{n!}$$

for each $n \in I\!N_0$,

(iii)   for $0 \leq t_0 < t_1 < \ldots < t_l < \infty$,

$$\{N_{t_0}; N_{t_k} - N_{t_{k-1}}, k = 1, \ldots, l\}$$

is a set of independent r.v.'s.

Any Poisson process has a version with right continuous paths (this is an easy consequence of Theorem 3 in Chung [12, p. 29]). We shall always use this version. Almost surely, the paths of a Poisson process are constant except for upward jumps of size one, of which there are finitely many in each bounded time interval, but infinitely many in $[0, \infty)$. The times between successive jumps are independent and exponentially distributed with parameter α. Thus, if T_n is the time between the nth and $(n + 1)^{\text{st}}$ jump, then $P(T_n > t) = e^{-\alpha t}$ for each t. A typical sample path might look like that in Figure 1.1.

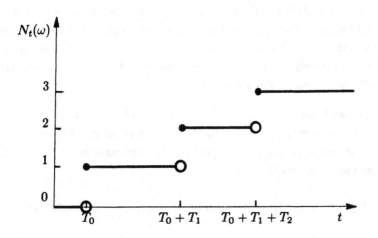

Figure 1.1.

For a Poisson process N, there is an associated standard filtration $\{\mathcal{F}_t\}$ defined by $\mathcal{F}_t = \sigma\{N_s, 0 \leq s \leq t\}^{\sim}$ for $t \in \mathbb{R}_+$, where the inclusion of the P-null sets of $\mathcal{F}$ in $\mathcal{F}_t$ ensures that $\mathcal{F}_t = \mathcal{F}_{t+}$. We shall always use this filtration when dealing with a Poisson process.

Brownian motion. A process $B = \{B_t, t \in \mathbb{R}_+\}$ is called a *Brownian motion* in $\mathbb{R}$ if it has the following properties:

(i) for $0 \leq s < t < \infty$, $B_t - B_s$ is a normally distributed r.v. with mean zero and variance $t - s$;

(ii) for $0 \leq t_0 < t_1 < \ldots < t_l < \infty$,

$$\{B_{t_0}; \ B_{t_k} - B_{t_{k-1}}, k = 1, \ldots, l\}$$

is a set of independent r.v.'s.

A *Brownian motion in* $\mathbb{R}^d$ is a d-tuple

$$\mathbf{B} = \{\mathbf{B}_t = (B_t^1, B_t^2, \ldots, B_t^d), t \in \mathbb{R}_+\}$$

where each $B^i = \{B_t^i, t \in \mathbb{R}_+\}, i = 1, 2, \ldots, d$, is a Brownian motion in $\mathbb{R}$ and the B^i's are independent. When the dimension of a Brownian motion is unimportant or understood, we suppress the qualifier "in $\mathbb{R}^d$". We use P^x and E^x respectively to denote the probability and expectation associated with a Brownian motion $\mathbf{B}$ for which $\mathbf{B}_0 = x$ a.s.

It follows from property (i) of each of the independent components of $\mathbf{B}$ that the distribution of $\mathbf{B}_t - \mathbf{B}_s$ depends only on the difference $t - s$. This is referred to as *temporal homogeneity*. Furthermore, if $\mathbf{B}_0 = x$ a.s., then the transition probability

$$P_t(x, A) \equiv P^x(\mathbf{B}_t \in A) = (2\pi t)^{-\frac{1}{2}d} \int_A \exp\left(-|x - y|^2/2t\right) dy$$

for any $t > 0$, $x \in \mathbb{R}^d$, and Borel set A in $\mathbb{R}^d$. Consequently,

$$P_t(x_0 + x, x_0 + A) = P_t(x, A) \text{ for any } x_0 \in \mathbb{R}^d.$$

This is called *spatial homogeneity*. Property (ii) above, which also holds when B is replaced by $\mathbf{B}$ for $d > 1$, is referred to as the *independence of increments* property of Brownian motion.

It is well known (see for example Chung [12, p. 145]) that any Brownian motion has a version with continuous paths. We shall always use such a version. Just as for a Poisson process, there is a standard filtration associated with a Brownian motion $\mathbf{B}$. It is defined by $\mathcal{F}_t = \sigma\{\mathbf{B}_s, 0 \leq s \leq t\}^\sim$ for each t. When dealing with a Brownian motion, unless specifically indicated otherwise, we shall use this filtration and define optionality with respect to it.

One of the basic properties of Brownian motion is the *strong Markov property*. Loosely speaking this says that given the history of a Brownian motion $\mathbf{B}$ up to some finite optional time τ, the behavior of $\mathbf{B}$ after that time depends only on τ and the state $\mathbf{B}_\tau$ of $\mathbf{B}$ at time τ. More precisely, if

$f : I\!R^d \to I\!R$ is a bounded Borel measurable function and τ is an optional time, then

$$E^x(1_{\{\tau<\infty\}} f(\mathbf{B}_{\tau+t}) \mid \mathcal{F}_\tau) = 1_{\{\tau<\infty\}} E^{\mathbf{B}_\tau} (f(\mathbf{B}_t)) .$$

If A is any Borel set in $I\!R^d$, then

$$\tau_A \equiv \inf\{t > 0 : B_t \notin A\}$$

is an optional time. See Chung [12, p. 92].

1.9 Martingales

In this section and the next, $\{\mathcal{F}_t, t \in I\!R_+\}$ is a filtration and optional means optional with respect to $\{\mathcal{F}_t\}$.

A collection $M = \{M_t, \mathcal{F}_t, t \in I\!R_+\}$ is called a *martingale* iff

(i) $M_t \in L^1$ for each t,

(ii) $M_s = E(M_t \mid \mathcal{F}_s)$ for all $s < t$.

Equality in (ii) means that M_s is a member of the equivalence class of $\mathcal{F}_s$-measurable r.v.'s which represent the conditional expectation of M_t given $\mathcal{F}_s$. We call M a *submartingale* iff the "=" in (ii) is replaced by "$\leq$" and a *supermartingale* iff it is replaced by "$\geq$". This condition is referred to as the (sub/super) martingale property when M is a (sub/super) martingale. We shall often omit the filtration and/or $t \in I\!R_+$ from the notation when it is understood, e.g., $M = \{M_t\}$ is a martingale means $M = \{M_t, \mathcal{F}_t, t \in I\!R_+\}$ is a martingale.

For $p \in [1, \infty)$, M is called an L^p-*martingale* iff it is a martingale and $M_t \in L^p$ for each t. If $\sup_{t \in I\!R_+} E(|M_t|^p) < \infty$, we say M is L^p-*bounded*.

The martingale property is preserved by L^p-limits when $\mathcal{F}_0$ contains all of the P-null sets in $\mathcal{F}$. More precisely we have the following.

Proposition 1.3. *Let $p \in [1, \infty)$. Suppose $\{M_t^n, \mathcal{F}_t, t \in I\!R_+\}$ is an L^p-martingale for each $n \in I\!N$, and for each t, M_t^n converges in L^p to M_t as*

$n \to \infty$. If $\mathcal{F}_0$ contains all of the P-null sets in $\mathcal{F}$, then $\{M_t, \mathcal{F}_t, t \in I\!\!R_+\}$ is an L^p-martingale.

Proof. It suffices to verify condition (ii) in the definition of a martingale. Fix $s < t$ in $I\!\!R_+$. For each n,

$$(1.2) \qquad\qquad M_s^n = E(M_t^n \mid \mathcal{F}_s).$$

The left side above converges in L^p to M_s, by hypothesis, and by Proposition 1.2, the right side converges to $E(M_t \mid \mathcal{F}_s)$ in L^p. Hence

$$M_s = E(M_t \mid \mathcal{F}_s) \text{ a.s.}$$

If $\mathcal{F}_0$ contains all of the P-null sets in $\mathcal{F}$, it follows that $M_s \in \mathcal{F}_s$ and then the a.s. above may be removed. ∎

The definition given above is of a martingale with continuous parameter $t \in I\!\!R_+$. Occasionally we shall refer to martingales with the parameter t restricted to a subset of $I\!\!R_+$ (e.g., a sub-interval or a discrete set of points). They are defined by restricting the defining conditions (i) and (ii) to that subset. The theory of continuous parameter martingales is discussed in Section 1.4 of Chung [12], which builds on the theory of discrete parameter martingales treated in Chapter 9 of Chung [11].

If $M = \{M_t, \mathcal{F}_t, t \in I\!\!R_+\}$ is a martingale, then for each constant $T \in I\!\!R_+$ it is easy to verify that $M^T = \{M_{t \wedge T}, \mathcal{F}_t, t \in I\!\!R_+\}$ is a martingale, and since $M_{t \wedge T} = E(M_T \mid \mathcal{F}_t)$, it follows that M^T is uniformly integrable. When T is replaced by a bounded optional time τ, a similar result holds (see Corollary 1.7(ii) in the following), provided M has right continuous paths and the filtration is standard. Indeed, many useful theorems for continuous parameter martingales require these hypotheses. We therefore adopt the following definition.

A martingale $M = \{M_t, \mathcal{F}_t, t \in I\!\!R_+\}$ is called *(right) continuous* iff

(i) $\{\mathcal{F}_t, t \in I\!\!R_+\}$ is a standard filtration, and

(ii) $\{M_t, t \in I\!\!R_+\}$ has all paths (right) continuous.

Example 1. Let $N = \{N_t, t \in \mathbb{R}_+\}$ be a Poisson process with parameter $\alpha > 0$ and $\{\mathcal{F}_t\}$ be the associated standard filtration. Define $M_t = N_t - \alpha t$. Then $\{M_t, \mathcal{F}_t, t \in \mathbb{R}_+\}$ and $\{M_t^2 - \alpha t, \mathcal{F}_t, t \in \mathbb{R}_+\}$ are right continuous L^p-martingales for each $p \in [1, \infty)$.

Example 2. Let $B = \{B_t, t \in \mathbb{R}_+\}$ be a Brownian motion in $\mathbb{R}$ with $B_0 \in L^p$ for some $p \in [1, \infty)$ and let $\{\mathcal{F}_t\}$ be the standard filtration associated with B. Then $\{B_t, \mathcal{F}_t, t \in \mathbb{R}_+\}$ is a continuous L^p-martingale. Moreover, if $p \geq 2$, then $\{B_t^2 - t, \mathcal{F}_t, t \in \mathbb{R}_+\}$ is a continuous $L^{p/2}$-martingale.

If M is a martingale and condition (i) above is satisfied, then there is a right continuous version of M (see Theorem 3 in Chung [12, p. 29]). However, there need not be a continuous version of it.

Some basic results from martingale theory are reviewed below.

Theorem 1.4. Let $p \in [1, \infty)$ and M be a right continuous L^p-martingale. Then for each t and $c \geq 0$,

$$(1.3) \qquad c^p P(\sup_{0 \leq s \leq t} |M_s| \geq c) \leq E(|M_t|^p ; \sup_{0 \leq s \leq t} |M_s| \geq c).$$

If $p > 1$, then for each t, $\sup_{0 \leq s \leq t} |M_s| \in L^p$ and

$$(1.4) \qquad \| \sup_{0 \leq s \leq t} |M_s| \|_p \leq q \|M_t\|_p$$

where $1/p + 1/q = 1$.

Proof. Inequality (1.3) follows by applying the discrete parameter Theorem 9.4.1 of Chung [11] to the submartingale $|M|^p$ evaluated at finitely many time points, and then taking the limit as these points become dense in $[0, t]$. In a similar fashion, (1.4) follows by applying Theorem 9.5.4 of Chung [11] to the positive submartingale $|M|$. ∎

Inequality (1.4) will be called *Doob's inequality*. As in the discrete parameter case, it is replaced by a more complicated result when $p = 1$ (see Chung [11, p. 355]).

Theorem 1.5. *(Martingale convergence theorem).* Let $p \in [1, \infty)$ and M be a right continuous L^p-bounded martingale. Then there is a r.v. $M_\infty \in L^p$ such that $\lim_{t \to \infty} M_t = M_\infty$ a.s. Furthermore, if either of the following conditions holds:

(i) $p = 1$ and $\{M_t, t \in I\!R_+\}$ is uniformly integrable, or

(ii) $p > 1$,

then $M_t \to M_\infty$ in L^p as $t \to \infty$, $\{M_t, \mathcal{F}_t, t \in [0, \infty]\}$ is an L^p-martingale where $\mathcal{F}_\infty = \bigvee_{t \in I\!R_+} \mathcal{F}_t$,

$$E\left(|M_\infty|^p\right) = \lim_{t \uparrow \infty} \uparrow E\left(|M_t|^p\right),$$

and (1.4) holds with $t = \infty$.

Proof. The existence of $M_\infty \in L^1$ such that $\lim_{t \to \infty} M_t = M_\infty$ a.s. follows from Corollary 2 in Chung [12, p. 27]. By Fatou's lemma we have:

$$(1.5) \qquad E\left(|M_\infty|^p\right) \leq \liminf_{n \to \infty} E\left(|M_n|^p\right) \leq \sup_{t \in I\!R_+} E\left(|M_t|^p\right) < \infty.$$

If either condition (i) or (ii) holds, then it follows from Proposition 1.1 that M_n converges to M_∞ in L^1 as $n \to \infty$. Now, for $t < n$ we have

$$M_t = E\left(M_n \mid \mathcal{F}_t\right),$$

and by letting $n \to \infty$ and using Proposition 1.2, we obtain

$$(1.6) \qquad\qquad M_t = E\left(M_\infty \mid \mathcal{F}_t\right).$$

Since $M_\infty = \lim_{t \uparrow \infty} M_t$ a.s. and $\mathcal{F}_\infty$ contains all of the P-null sets in $\mathcal{F}$, we have $M_\infty \in \mathcal{F}_\infty$. Thus, $\{M_t, \mathcal{F}_t, t \in [0, \infty]\}$ is an L^p-martingale. Hence, $\{|M_t|^p, \mathcal{F}_t, t \in [0, \infty]\}$ is a submartingale, and therefore

$$\lim_{t \uparrow \infty} \uparrow E\left(|M_t|^p\right) = \sup_{t \in I\!R_+} E\left(|M_t|^p\right) \leq E\left(|M_\infty|^p\right).$$

By combining this with (1.5), we see that the last inequality above is actually an equality. Then it follows (cf. Proposition 1.1) that $M_t \to M_\infty$ in L^p as $t \to \infty$. Inequality (1.4) with $t = \infty$ follows on letting $t \to \infty$ there.

∎

Theorem 1.6. *(Doob's Stopping Theorem).* Let $p \in [1, \infty)$ and M be a right continuous L^p-bounded martingale. If $p = 1$, suppose M is also uniformly integrable. Let M_∞ be as in Theorem 1.5. Suppose $\Gamma \subset I\!\!R_+$ and $\{\tau_t, t \in \Gamma\}$ is an increasing family of optional times. Then $\{M_{\tau_t}, \mathcal{F}_{\tau_t}, t \in \Gamma\}$ is an L^p-martingale and $\{|M_{\tau_t}|^p, t \in \Gamma\}$ is uniformly integrable.

Proof. By Theorem 1.5, $\{M_t, \mathcal{F}_t, t \in [0, \infty]\}$ is an L^p-martingale and (1.6) holds for all t. It follows from Theorem 4 of Chung [12, p. 30] that for any two optional times $\eta \leq \sigma$, M_η and M_σ are in L^1, and $M_\eta = E(M_\sigma \,|\, \mathcal{F}_\eta)$. By setting $\eta = \tau_s$ and $\sigma = \tau_t$ for $s < t$ in Γ, the martingale property follows for $\{M_{\tau_t}, \mathcal{F}_{\tau_t}, t \in \Gamma\}$. On the other hand, if we let $\eta = \tau_t$ for $t \in \Gamma$ and $\sigma = \infty$, then by Jensen's inequality for conditional expectations we have

$$|M_{\tau_t}|^p = |E(M_\infty \,|\, \mathcal{F}_{\tau_t})|^p \leq E(|M_\infty|^p \,|\, \mathcal{F}_{\tau_t}).$$

It follows from this that $M_{\tau_t} \in L^p$ for each $t \in \Gamma$, and $\{|M_{\tau_t}|^p, t \in \Gamma\}$ is uniformly integrable. ∎

Corollary 1.7. *Let $p \in [1, \infty)$ and M be a right continuous L^p-martingale.*

(i) *If $\Gamma \subset I\!\!R_+$ and $\{\tau_t, t \in \Gamma\}$ is an increasing family of optional times such that $\sup_{t \in \Gamma} \tau_t \leq T$ for some $T \in I\!\!R_+$, then $\{M_{\tau_t}, \mathcal{F}_{\tau_t}, t \in \Gamma\}$ is an L^p-martingale and $\{|M_{\tau_t}|^p, t \in \Gamma\}$ is uniformly integrable.*

(ii) *If τ is an optional time, then $\{M_{t \wedge \tau}, \mathcal{F}_t, t \in I\!\!R_+\}$ is an L^p-martingale. Moreover, if τ is bounded, then $\{|M_{t \wedge \tau}|^p, \ t \in I\!\!R_+\}$ is uniformly integrable.*

Proof. For part (i), since $M_{t \wedge T} = E(M_T \,|\, \mathcal{F}_t)$ and $M_T \in L^p$, it follows that $\{M_{t \wedge T}, \mathcal{F}_t, t \in I\!\!R_+\}$ satisfies the hypotheses of Theorem 1.6. The conclusion in (i) then follows since $\tau_t \wedge T = \tau_t$.

For part (ii), apply (i) with $\tau_t = t \wedge \tau$ for $t \in \Gamma = [0, T]$ and fixed $T \in I\!\!R_+$. Since T is arbitrary, we conclude that $\{M_{t \wedge \tau}, \mathcal{F}_{t \wedge \tau}, t \in I\!\!R_+\}$ is an L^p-martingale. We leave it as an exercise to verify that $\mathcal{F}_{t \wedge \tau}$ may be replaced by $\mathcal{F}_t$ here. If τ is bounded by T, the uniform integrability follows from that on $[0, T]$. ∎

1.10 Local Martingales

For $p \in [1, \infty)$, a collection $M = \{M_t, \mathcal{F}_t, t \in \mathbb{R}_+\}$ is called a *local L^p-martingale* iff

(i) M_0 is an $\mathcal{F}_0$-measurable r.v.,

(ii)   there is a sequence $\{\tau_k, k \in \mathbb{N}\}$ of optional times such that $\tau_k \uparrow \infty$ a.s. and for each k,

(1.7) $$M^k = \{M_{t \wedge \tau_k} - M_0, \mathcal{F}_t, t \in \mathbb{R}_+\}$$

is an L^p-martingale.

The sequence $\{\tau_k\}$ is called a *localizing sequence* for M. When $p = 1$, we omit the qualifier "L^p". If $M_t = M_0 \in \mathcal{F}_0$ for all t, then M is a local martingale according to the above definition. A less trivial example is a Brownian motion in $\mathbb{R}$ with an arbitrary initial r.v. and the usual filtration $\mathcal{F}_t = \sigma\{B_s, 0 \leq s \leq t\}^\sim$. The above definition is motivated by such examples, where the initial r.v. M_0 is not subject to any integrability conditions.

We shall usually omit the filtration $\{\mathcal{F}_t\}$ from the notation for a local martingale.

In their definition of a local martingale, Dellacherie and Meyer [24, p. 94] require M^k to be uniformly integrable for each k. This is not more restrictive than (ii) above, for if $\{M_{t \wedge \tau_k} - M_0, t \in \mathbb{R}_+\}$ is a martingale, then $\{M_{t \wedge k \wedge \tau_k} - M_0, t \in \mathbb{R}_+\}$ is a uniformly integrable martingale.

A local L^p-martingale $M = \{M_t, \mathcal{F}_t, t \in \mathbb{R}_+\}$ is called *(right) continuous* iff

(i) $\{\mathcal{F}_t, t \in \mathbb{R}_+\}$ is a standard filtration, and

(ii) M has all paths (right) continuous.

Clearly an L^p-martingale ($p \in [1, \infty)$) is a local L^p-martingale. Conditions implying the converse are given below.

Proposition 1.8. *Let $p \in [1, \infty)$ and M be a local L^p-martingale with a localizing sequence $\{\tau_k\}$. If for each $t \geq 0$ we have*

(1.8) $\{|M_{t \wedge \tau_k}|^p, k \in I\!N\}$ *is uniformly integrable,*

then M is an L^p-martingale. The converse is true provided that M is right continuous.

Proof. Suppose (1.8) is true. Then for $t = 0$, we have $|M_0| \in L^p$, and consequently $\{M_0, \mathcal{F}_t, t \in I\!R_+\}$ is an L^p-martingale. It follows by addition with (1.7) that $\{M_{t \wedge \tau_k}, \mathcal{F}_t, t \in I\!R_+\}$ is an L^p-martingale. Now $\lim_{k \to \infty} M_{t \wedge \tau_k} = M_t$ a.s., and the uniform integrability (1.8) implies that we also have convergence in L^p, by Proposition 1.1. Since $M_t \in \mathcal{F}_t$ for each t, it then follows similarly to Proposition 1.3 that $\{M_t, \mathcal{F}_t, t \in I\!R_+\}$ is an L^p-martingale.

If M is a right continuous L^p-martingale, then it follows from Corollary 1.7(i) that $\{|M_{t \wedge \tau_k}|^p, k \in I\!N\}$ is uniformly integrable for each fixed t. ∎

If M is a *continuous* local martingale, then there is a natural choice of localizing sequence for M which shows that M is a local L^p-martingale for any $p \in [1, \infty)$. This sequence is exhibited below.

Proposition 1.9. *Suppose M is a continuous local martingale and let $\tau_k = \inf\{t > 0 : |M_t - M_0| > k\}$ for each $k \in I\!N$. Then, for each $p \in [1, \infty)$, M is a local L^p-martingale and $\{\tau_k\}$ is a localizing sequence for it.*

Proof. Let $\{\sigma_n\}$ be a localizing sequence for M, so that $\{M_{t \wedge \sigma_n} - M_0, t \in I\!R_+\}$ is a continuous martingale. Then by Corollary 1.7(ii),

$$\{M_{t \wedge \tau_k \wedge \sigma_n} - M_0, t \in I\!R_+\}$$

is a martingale for each k and n. By the definition of τ_k, the above is bounded by k. Hence $M^k = \{M_{t \wedge \tau_k} - M_0, t \in I\!R_+\}$ is a martingale by Proposition 1.8 (with τ_k there replaced by σ_n). Thus, $\{\tau_k\}$ is a localizing sequence for M, such that for each k, M^k is bounded and hence in L^p for each $p \in [1, \infty)$. ∎

We end this discussion of local martingales with an example of an L^2-bounded local martingale which is not a martingale.

Example.   Let $\mathbf{B} = \{\mathbf{B}_t, t \in I\!\!R_+\}$ denote a Brownian motion in $I\!\!R^3$ satisfying $\mathbf{B}_0 \neq 0$. Let $h : I\!\!R^3 \backslash \{0\} \to I\!\!R$ be defined by $h(x) = |x|^{-1}$ for $x \in I\!\!R^3 \backslash \{0\}$. For each $k \in I\!\!N$, let $\tau_k = \inf\{t \geq 0 : |\mathbf{B}_t| \leq k^{-1}\}$. Then $\{\tau_k\}$ is an increasing sequence of optional times with respect to the filtration $\{\mathcal{F}_t\}$ associated with $\mathbf{B}$, and $\tau_k \uparrow \infty$ a.s. since

$$P\{\mathbf{B}_t = 0 \text{ for some } t \geq 0\} = 0.$$

The function h is harmonic in $I\!\!R^3 \backslash \{0\}$, which contains

$$D_k = \{x : |x| > k^{-1}\} \quad \text{for each} \quad k.$$

Define a function g_k on the closure $\overline{D_k}$ of D_k by

$$g_k(x) = E^x\left(h(\mathbf{B}_{\tau_k}); \tau_k < \infty\right) \quad \text{for each } x \in \overline{D_k},$$

where E^x denotes the expectation given $\mathbf{B}_0 = x$ a.s. By the strong Markov property and spherical symmetry of $\mathbf{B}$, g_k possesses the mean-value property that its average value over the surface of any sufficiently small ball about $x \in D_k$ equals its value at x. It follows that g_k is harmonic in D_k, and it can be shown to be continuous in $\overline{D_k}$ with boundary values on $\partial D_k \equiv \{x : |x| = k^{-1}\}$ equal to those of h (see Chung [12, §4.4]). Moreover, since Brownian motion is transient in three dimensions, $P^x(\tau_k < \infty) \to 0$ as $|x| \to \infty$, and h is bounded on ∂D_k. It follows that $g_k(x) \to 0$ as $|x| \to \infty$. Note that $h(x) \to 0$ as $|x| \to \infty$. Then, by the maximum principle for harmonic functions, $g_k = h$ in $\overline{D_k}$ for all k. For $k \in I\!\!N$ and $x \in D_k$, we have for each fixed t:

$$E^x\left(h(\mathbf{B}_{\tau_k}) \mid \mathcal{F}_t\right) = 1_{\{\tau_k \leq t\}} h(\mathbf{B}_{t \wedge \tau_k}) + 1_{\{\tau_k > t\}} E^x\left(h(\mathbf{B}_{\tau_k}) \mid \mathcal{F}_t\right).$$

By the strong Markov property, on $\{\tau_k > t\}$ we have

$$E^x\left(h(\mathbf{B}_{\tau_k}) \mid \mathcal{F}_t\right) = E^{\mathbf{B}_t}\left(h(\mathbf{B}_{\tau_k})\right) = g_k(\mathbf{B}_t).$$

By combining the above, since $g_k = h$ in $\overline{D_k}$, we have

$$E^x\left(h(\mathbf{B}_{\tau_k}) \mid \mathcal{F}_t\right) = h(\mathbf{B}_{t \wedge \tau_k}).$$

Suppose $\mathbf{B}_0 \equiv x_0 \neq 0$. For all sufficiently large k, $x_0 \in D_k$ and then by the above we have $\{h\left(\mathbf{B}_{t \wedge \tau_k}\right), t \in I\!\!R_+\}$ is a bounded martingale. Since $\tau_k \uparrow \infty$ as $k \uparrow \infty$, it follows that $\{h(\mathbf{B}_t), t \in I\!\!R_+\}$ is a local martingale. But it is not a martingale because

$$E^{x_0}\left(h(\mathbf{B}_t)\right) \neq E^{x_0}\left(h(\mathbf{B}_0)\right) = |x_0|^{-1} \text{ for all sufficiently large } t,$$

by the following calculation. For $t > 0$ and $R > 2\,|x_0|$,

$$E^{x_0}\left(h(\mathbf{B}_t)\right) = \frac{1}{(2\pi t)^{\frac{3}{2}}} \int\limits_{I\!\!R^3} |y|^{-1}\, e^{-|y - x_0|^2/2t}\, dy$$

$$\leq \frac{1}{(2\pi t)^{\frac{3}{2}}} \left\{ \int\limits_{|y| \leq R} |y|^{-1}\, dy + \int\limits_{|y| > R} |y|^{-1}\, e^{-|y|^2/8t}\, dy \right\}$$

$$\leq \frac{C_1 R^2}{(2\pi t)^{\frac{3}{2}}} + \frac{C_2}{R}.$$

Here $y \in I\!\!R^3$, and C_1, C_2 are constants independent of t and R. By letting $t \to \infty$ and then $R \to \infty$, it follows that $\lim_{t \to \infty} E^{x_0}\left(h(\mathbf{B}_t)\right) = 0$.

A similar calculation shows that $h(\mathbf{B}_t) \in L^2$ for each t and more refined estimates yield $\sup_{t \geq 0} E^{x_0}\left\{(h(\mathbf{B}_t))^2\right\} < \infty$.

Convention. For the rest of this book, $\{\mathcal{F}_t, t \in I\!\!R_+\}$ will denote a *standard* filtration and, unless otherwise specified, optional times and adapted processes will be with respect to $\{\mathcal{F}_t\}$.

1.11 Exercises

1. If M is a positive local martingale satisfying $E(M_0) < \infty$, then M is a supermartingale.
Hint: Use Fatou's lemma for conditional expectations.

For Exercises 2–3 below,   let $\{N_t, t \in \mathbb{R}_+\}$ be a Poisson process with parameter $\alpha > 0$ and let $\{\mathcal{F}_t, t \geq 0\}$ be the associated standard filtration as defined in Section 1.8.

2.   Define $M_t = N_t - \alpha t$ for all $t \in \mathbb{R}_+$. Prove the claims in Example 1 of Section 1.9 that $\{M_t, \mathcal{F}_t, t \in \mathbb{R}_+\}$ and $\{M_t^2 - \alpha t, \mathcal{F}_t, t \in \mathbb{R}_+\}$ are L^p-martingales for each $p \in [1, \infty)$. Is either of them uniformly integrable? *Hint:* Write $M_t^2 - M_s^2 = (M_t - M_s)^2 + 2M_s(M_t - M_s)$.

3.   Let $T_n = \inf\{t \geq 0 : N_t \geq n\}$. Prove that $E(T_n) = n/\alpha$.

4.   Verify the claims made in Example 2 of Section 1.9 that when $B_0 \in L^p$ for some $p \geq 1$, $\{B_t, \mathcal{F}_t, t \in \mathbb{R}_+\}$ is an L^p-martingale and $\{B_t^2 - t, \mathcal{F}_t, t \in \mathbb{R}_+\}$ is an $L^{p/2}$-martingale for $p \geq 2$. You may use Proposition 1.8 if you wish.
Hint: Write $B_t = (B_t - B_0) + B_0$ and $B_t^2 - B_s^2 = (B_t - B_s)^2 + 2B_s(B_t - B_s)$.

5.   Let τ be an exponentially distributed random variable with parameter $\lambda > 0$, so that $P(\tau > t) = e^{-\lambda t}$ for all $t \geq 0$. Define $X_t = e^{\lambda t} 1_{\{t < \tau\}}$ for all $t \geq 0$. Let $\mathcal{F}_t = \sigma\{X_s : 0 \leq s \leq t\}$. Prove that $\{X_t, \mathcal{F}_t, t \geq 0\}$ is a positive martingale, but it is not uniformly integrable. Define $X_{0-} = X_0$ and show that $X_{\tau-} \notin L^1$.

2

DEFINITION OF THE STOCHASTIC INTEGRAL

2.1 Introduction

In this chapter, we shall define stochastic integrals of the form $\int_{[0,t]} X \, dM$ where M is a right continuous local L^2-martingale and X is a process satisfying certain measurability and integrability assumptions, such that the family of stochastic integrals $\{\int_{[0,t]} X \, dM, t \in \mathbb{R}_+\}$ is a right continuous local L^2-martingale. For certain M and X, the integral can be defined path-by-path. For instance, if M is a right continuous local L^2-martingale whose paths are locally of bounded variation, and X is a continuous adapted process, then $\int_{[0,t]} X_s(\omega) \, dM_s(\omega)$ is well-defined as a Riemann-Stieltjes integral for each t and ω, namely by the limit as $n \to \infty$ of

$$\sum_{k=0}^{[2^n t]} X_{k2^{-n}}(\omega) \left(M_{(k+1)2^{-n}}(\omega) - M_{k2^{-n}}(\omega) \right).$$

The standard example of this path-by-path integral is obtained by setting $M_t = N_t - \alpha t$ where N is a Poisson process with parameter $\alpha > 0$. In this case, for any continuous adapted process X we have

$$\int_{[0,t]} X_s(\omega) \, dM_s(\omega) = \sum_{k=1}^{\infty} 1_{\{\tau_k \leq t\}} X_{\tau_k}(\omega) - \alpha \int_0^t X_s(\omega) \, ds,$$

where τ_k is the time of the kth jump of N, and a.s. for each fixed t the sum on the right is of finitely many non-zero terms because almost surely there are only finitely many jumps of N in $[0, t]$.

The stochastic integral defined in the sequel is valid even when M does not have paths which are locally of bounded variation. Any non-constant continuous local martingale is such an M; the canonical example is a Brownian motion B in $\mathbb{R}$. Even the simple integral $\int_{[0,t]} B \, dB$ cannot be defined path-by-path in the Stieltjes sense, because almost every path of a Brownian motion is of unbounded variation on each time interval (see Freedman [33, p. 49]). In fact, the stochastic integral developed here, known as the Itô integral when M is a Brownian motion, is not defined path-by-path but via an isometry between a space of processes X that are square integrable with respect to a measure induced by M, and a space of square integrable stochastic integrals $\int X \, dM$.

As a guide to the reader, we provide the following outline of the several stages in the definition of the stochastic integral.

The measurability conditions on X will be specified first. In doing this, we adopt the modern view of X as a function on $\mathbb{R}_+ \times \Omega$ and require it to be measurable with respect to a σ-field $\mathcal{P}$ generated by a simple class $\mathcal{R}$ of "predictable rectangles." Although this definition of the measurable integrands may not be the most obvious one, it is convenient for a streamlined development of the integral. Moreover, we shall prove in Theorem 3.1 that the class of $\mathcal{P}$-measurable functions includes all of the left continuous adapted processes.

After a discussion of the σ-field $\mathcal{P}$, we shall consider the case where M is a right continuous L^2-martingale. A measure μ_M associated with M will be defined on $\mathcal{P}$ and then we shall define the integral $\int_{[0,t]} X \, dM$ in the following three steps.

(i) $\int X \, dM$ will be defined for any $\mathcal{R}$-simple process X in such a way that the following isometry holds:

$$E\left\{\left(\int X\,dM\right)^2\right\} = \int_{I\!R_+\times\Omega} (X)^2\,d\mu_M.$$

(ii)   This isometry will then be used to extend the definition of $\int X\,dM$
to any $X \in \mathcal{L}^2 \equiv L^2(I\!R_+ \times \Omega, \mathcal{P}, \mu_M)$.

(iii) For any process X satisfying $1_{[0,t]}X \in \mathcal{L}^2$ for each $t \in I\!R_+$, it will
be shown that there is a version of $\{\int 1_{[0,t]}X\,dM, t \in I\!R_+\}$ which is
a right continuous L^2-martingale, to be denoted by $\{\int_{[0,t]} X\,dM, t \in I\!R_+\}$.

Finally, the extension to the case where M is a right continuous local
L^2-martingale and X is "locally" in $\mathcal{L}^2$ will be achieved using a sequence of
optional times tending to ∞. The above definition of the stochastic integral
will apply to the processes obtained by stopping $M - M_0$ and X at any one
of these times, and then the integral for M and X will be defined as the
almost sure limit of these integrals, as the optional times tend to ∞.

We now begin the above program with the definition of the σ-field $\mathcal{P}$.

2.2 Predictable Sets and Processes

The family of subsets of $I\!R_+ \times \Omega$ containing all sets of the form $\{0\} \times F_0$
and $(s, t] \times F$, where $F_0 \in \mathcal{F}_0$ and $F \in \mathcal{F}_s$ for $s < t$ in $I\!R_+$, is called the
class of predictable rectangles and we denote it by $\mathcal{R}$. The (Boolean) ring $\mathcal{A}$
generated by $\mathcal{R}$ is the smallest family of subsets of $I\!R_+ \times \Omega$ which contains
$\mathcal{R}$ and is such that if A_1 and A_2 are in the ring, then so too are their union
$A_1 \cup A_2$ and difference $A_1 \backslash A_2$. Then $A_1 \cap A_2$ is also in $\mathcal{A}$. Indeed, it can be
verified that the ring $\mathcal{A}$ consists of the empty set $\emptyset$ and all finite unions of
disjoint rectangles in $\mathcal{R}$. The σ-field $\mathcal{P}$ of subsets of $I\!R_+ \times \Omega$ generated by $\mathcal{R}$
is called the *predictable σ-field* and sets in $\mathcal{P}$ are called *predictable (sets)*.
A function $X : I\!R_+ \times \Omega \to I\!R$ is called predictable if X is $\mathcal{P}$-measurable.
This is denoted by $X \in \mathcal{P}$. If A is a set in $\mathcal{R}$, then $1_A(t, \cdot)$ is $\mathcal{F}_t$-measurable
for each t. Consequently, 1_A is an adapted process. It follows by forming

finite linear combinations that the same is true for any A in $\mathcal{A}$. Then by a monotone class theorem (see Section 1.2), any real-valued $\mathcal{P}$-measurable function is adapted. A real-valued $\mathcal{P}$-measurable function will be referred to as a predictable *process*.

Remark. In systematic studies of the theory of processes, it seems more natural to consider the σ-field $\mathcal{P}$ and predictable processes as defined on $(0, \infty) \times \Omega$. However, we find it convenient to have all processes defined at time zero. The consequence, which is of more logical than substantial significance, is that time zero and sets like $\{0\} \times F_0$ sometimes require slightly different treatment.

It is shown below that for any optional time τ,

$$[0, \tau] = \{(t, \omega) \in \mathbb{R}_+ \times \Omega : 0 \le t \le \tau(\omega)\}$$

is a predictable set. Such "intervals" play an important role in the final extension phase of the definition of the stochastic integral.

2.3 Stochastic Intervals

For optional times η and τ, the set

$$[\eta, \tau] = \{(t, \omega) \in \mathbb{R}_+ \times \Omega : \eta(\omega) \le t \le \tau(\omega)\}$$

is called a stochastic interval. Three other stochastic intervals $(\eta, \tau], (\eta, \tau)$, and $[\eta, \tau)$, with left end-point η and right end-point τ are defined similarly. The term *stochastic interval* will refer to any of these four kinds of intervals where η and τ are any optional times. Note that stochastic intervals are subsets of $\mathbb{R}_+ \times \Omega$ not $\overline{\mathbb{R}}_+ \times \Omega$; consequently (∞, ω) is never a member of such a set, even if $\tau(\omega) = \infty$. Also, we have not specified that $\eta \le \tau$, but by definition the intersection of $[\eta, \tau]$ with $\mathbb{R}_+ \times \{\omega : \eta > \tau\}$ is the empty set. If $s, t \in \mathbb{R}_+$, then $[s, t], (s, t], [s, t)$ and (s, t), may be interpreted as real or stochastic intervals. It will usually be clear from the context which interpretation is meant. For example, in equation (2.10), $1_{[0, t]}$ means the indicator function of the stochastic interval $[0, t] \times \Omega$.

The σ-field of subsets of $I\!R_+ \times \Omega$ generated by the class of stochastic intervals is called the *optional* σ-field and is denoted by $\mathcal{O}$. The graph of an optional time τ, denoted by

$$[\tau] = [0, \tau]\backslash[0, \tau) = \{(t, \omega) \in I\!R_+ \times \Omega : \tau(\omega) = t\},$$

is in $\mathcal{O}$. A function $X : I\!R_+ \times \Omega \to I\!R$ will be called *optional* iff X is $\mathcal{O}$-measurable. If A is a stochastic interval, then $1_A(t, \cdot)$ is $\mathcal{F}_t$-measurable for each t, by the optionality of the end-points of A. Then it follows as for predictable functions that any optional function is an adapted process, and we shall refer to it as an optional process.

We now investigate the relationship between $\mathcal{P}$ and $\mathcal{O}$. Each predictable rectangle of the form $(s, t] \times F$ where $F \in \mathcal{F}_s$ and $s < t$ in $I\!R_+$, is a stochastic interval of the form $(\eta, \tau]$ with $\eta \equiv s$, $\tau = s$ on $\Omega\backslash F$ and $\tau = t$ on F. Also, for $F_0 \in \mathcal{F}_0$, $\{0\} \times F_0 = \bigcap_n [0, \tau_n)$ where

$$\tau_n = \begin{cases} \frac{1}{n} & \text{on } F_0 \\ 0 & \text{on } \Omega\backslash F_0 \end{cases}$$

is optional for each n. It follows that $\mathcal{R} \subset \mathcal{O}$ and hence, since $\mathcal{R}$ generates $\mathcal{P}$, we have $\mathcal{P} \subset \mathcal{O}$. In the following lemma we show that certain types of stochastic intervals are predictable.

Lemma 2.1. *Stochastic intervals of the form $[0, \tau]$ and $(\eta, \tau]$ are predictable.*

Proof. Since $(\eta, \tau] = [0, \tau]\backslash[0, \eta]$, it suffices to prove that a stochastic interval of the form $[0, \tau]$ is predictable. For this we use a standard approximation of τ by a decreasing sequence $\{\tau_n\}$ of countably valued optional times, defined by $\tau_n = 2^{-n}[2^n\tau + 1]$. Since $\tau_n \downarrow \tau$, we have $[0, \tau] = \bigcap_n [0, \tau_n]$. For each n,

$$[0, \tau_n] = (\{0\} \times \Omega) \cup \left(\bigcup_{k \in I\!N_0} (k2^{-n}, (k+1)2^{-n}] \times \{\tau \geq k2^{-n}\} \right).$$

Here $\{\tau \geq k2^{-n}\} = \Omega\backslash\{\tau < k2^{-n}\} \in \mathcal{F}_{k2^{-n}}$, since τ is optional. It follows that $[0, \tau] \in \mathcal{P}$. ∎

Stochastic intervals, other than those mentioned in the preceding lemma, are not in general predictable without further restriction on the end-points. An $\mathcal{F}$-measurable function $\tau : \Omega \to \overline{I\!\!R}_+$ is called a *predictable time* (or simply predictable) if there is a sequence of optional times $\{\tau_n\}$ which increases to τ such that each τ_n is *strictly* less than τ on $\{\tau \neq 0\}$. Such a sequence $\{\tau_n\}$ is called an *announcing sequence* for τ. It is easily verified that a predictable time is an optional time and as a partial converse, if τ is optional then $\tau + t$ is predictable for each constant $t > 0$. Intuitively speaking, if $\tau > 0$ is the first time some random event occurs, then τ is predictable if this event cannot take us by surprise because we are forewarned by a sequence of prior events, occurring at times τ_n. A very simple example of a predictable time is

$$0_{F_0} = \begin{cases} 0 & \text{on } F_0 \\ \infty & \text{on } F_0^c , \end{cases}$$

where $F_0 \in \mathcal{F}_0$. An announcing sequence for 0_{F_0} is $\{0_{F_0} \wedge n,\ n = 1, 2, \ldots\}$. An example of a non-predictable optional time is the time at which the first jump of a Poisson process occurs.

Parts (iii) and (iv) of the following lemma elucidate the reason for the names of the predictable and optional σ-fields.

Lemma 2.2.

(i)   If τ is a predictable time, then $[\tau, \infty)$ is predictable.

(ii)   All stochastic intervals of the following forms are predictable: $(\eta, \tau]$ where η and τ are optional, $[\eta, \tau]$ and (τ, η) where η is predictable and τ is optional, $[\eta, \tau)$ where η and τ are both predictable.

(iii) The predictable σ-field is generated by the class of stochastic intervals of the form $[\tau, \infty)$ where τ is a predictable time.

(iv)   The optional σ-field is generated by the class of stochastic intervals of the form $[\tau, \infty)$ where τ is an optional time.

Proof. To prove (i), suppose τ is a predictable time and $\{\tau_n\}$ is an announcing sequence for τ. Since $\tau_n \uparrow \tau$ and $\tau_n < \tau$ on $\{\tau \neq 0\}$, we have

$$[\tau, \infty) = (\{0\} \times \{\tau = 0\}) \cup \left(\bigcap_n (\tau_n, \infty) \right).$$

Here $\{\tau = 0\} \in \mathcal{F}_0$ and $(\tau_n, \infty) = (I\!R_+ \times \Omega) \backslash [0, \tau_n]$ is predictable for each n, by Lemma 2.1. Hence $[\tau, \infty)$ is predictable, proving (i).

For an optional time τ, $[0, \tau]$ is predictable by Lemma 2.1, and if τ is predictable, then $[0, \tau)$—the complement of $[\tau, \infty)$—is predictable by part (i) above. Since each of the four kinds of stochastic intervals in (ii) can be written as a difference of two intervals of the above kind, with η in place of τ in one of them, the result (ii) follows.

For the proof of (iii), let $\mathcal{Q}$ denote the σ-field generated by the class of stochastic intervals of the form $[\tau, \infty)$ where τ is predictable. By part (i), $\mathcal{Q} \subset \mathcal{P}$ and to show $\mathcal{P} \subset \mathcal{Q}$, it suffices to prove $\mathcal{R} \subset \mathcal{Q}$. For any optional time τ we have $[0, \tau] = \bigcap_n [0, \tau + \frac{1}{n})$. Here $\tau + \frac{1}{n}$ is predictable and therefore, by complementation, $[0, \tau + \frac{1}{n}) \in \mathcal{Q}$. Consequently, $[0, \tau] \in \mathcal{Q}$. A predictable rectangle $(s, t] \times F$ for $F \in \mathcal{F}_s$ and $s < t$, is a stochastic interval of the form $(\eta, \tau] = [0, \tau] \backslash [0, \eta]$ and is therefore in $\mathcal{Q}$. If $F_0 \in \mathcal{F}_0$, then since 0_{F_0} is a predictable time, we have $\{0\} \times F_0 = [0_{F_0}, \infty) \backslash (0, \infty) \in \mathcal{Q}$. Thus, $\mathcal{R} \subset \mathcal{Q}$ and hence (iii) is proved.

Since $\mathcal{O}$ is generated by the stochastic intervals, to prove (iv) it suffices to show that all stochastic intervals are contained in the σ-field $\mathcal{S}$ generated by the class of stochastic intervals of the form $[\tau, \infty)$. If τ is optional, then $\tau + \frac{1}{n}$ is optional for each n and hence $(\tau, \infty) = \bigcup_n [\tau + \frac{1}{n}, \infty)$ is in $\mathcal{S}$. Since the class consisting of the stochastic intervals of the form $[\tau, \infty)$ and (τ, ∞) generates all stochastic intervals by combinations of the operations of complementation and differencing, it follows that all stochastic intervals are in $\mathcal{S}$, as required. ∎

For $\tau : \Omega \to \overline{I\!R}_+$, we have by the above lemma:

(i)   if τ is predictable, then $[\tau, \infty)$ is predictable,

(ii)   if τ is optional, then $[\tau, \infty)$ is optional.

The converses of these results are also true. The converse of (ii) follows from the result proved earlier that if $[\tau, \infty)$ is an optional set, then $1_{[\tau, \infty)}$ is an adapted process. For the more difficult proof of the converse of (i), we refer the reader to Dellacherie and Meyer [23, IV-76]. (*Warning*: Dellacherie and Meyer use the conclusion of (i) as their definition of a predictable time and derive the existence of an announcing sequence from it). Alternative characterizations of the predictable and optional σ-fields to those of Lemma 2.2 will be given in Chapter 3.

We conclude this section with the following result which is well known to experts, especially those interested in applications to mathematical economics. It is also referred to later in Section 9.4. The proof given here was told to us by Michael Sharpe. The argument for the "if" part is similar to that in Chung-Walsh [19]; the proof of the "only if" part is standard.

Proposition. *Every optional time is predictable if and only if every (local) martingale (adapted to $\{\mathcal{F}_t\}$) has a continuous version.*

Proof. For the "if" part, suppose every martingale has a continuous version. Let τ be an optional time. We may assume that τ is bounded because $\tau \wedge n \uparrow \tau$ and the limit of an increasing sequence of predictable times is predictable. Consider the supermartingale Y defined by

$$Y_t = E[(\tau - t)^+ \mid \mathcal{F}_t] = E[\tau \mid \mathcal{F}_t] - \tau \wedge t.$$

Since $\{E[\tau \mid \mathcal{F}_t], t \geq 0\}$ is a martingale, by assumption we may choose a continuous version of it. Then Y has continuous sample paths.

We first prove that P-a.s., $Y_t = 0$ for all $t \geq \tau$ and $Y_t > 0$ for all $t < \tau$. For $t \geq \tau$, this follows from the fact that Y has continuous paths and

$$Y_t 1_{\{t \geq \tau\}} = E[(\tau - t)^+ 1_{\{t \geq \tau\}} \mid \mathcal{F}_t]$$
$$= E[0 \mid \mathcal{F}_t] = 0 \quad P\text{-a.s.}$$

To prove that $Y_t > 0$ for $t < \tau$, let $\sigma = \inf\{t \geq 0 : Y_t = 0\}$. Then by Doob's stopping theorem, P-a.s.,

$$Y_{\tau \wedge \sigma} = E[\tau \mid \mathcal{F}_{\tau \wedge \sigma}] - \tau \wedge \sigma = E[(\tau - \sigma)^+ \mid \mathcal{F}_{\tau \wedge \sigma}].$$

Now, by the result: $Y_\tau = 0$ P-a.s., the definition of σ implies that $Y_{\sigma \wedge \tau} = 0$ P-a.s. Thus, taking expectations in the above yields

$$0 = E[(\tau - \sigma)^+].$$

Hence $\tau \le \sigma$ P-a.s. and the desired property of Y follows. It then follows that $\tau_n \equiv \inf\{t \ge 0 : Y_t \le \frac{1}{n}\}$ is an announcing sequence for τ and hence τ is predictable.

For the "only if" part, suppose that every optional time is predictable. Let M be a local martingale. To prove M has a continuous version, it suffices by localization to consider the case where M is a uniformly integrable martingale. (Note that if M is a martingale, then $M_{\cdot \wedge n}$ is a uniformly integrable martingale for each positive integer n.) Then by Theorem 1.5, $M_\infty = \lim_{t \to \infty} M_t$ exists P-a.s. and $\{M_t, \mathcal{F}_t, t \in [0, \infty]\}$ is a martingale. Since every martingale has a version that is right continuous with finite left limits (see Chung [12, Section 1.4]), we may assume that M is such a version, and so M can only have jump discontinuities. For $\varepsilon > 0$, let $\tau \equiv \inf\{t \ge 0 : M_t - M_{t-} \ge \varepsilon\}$. Then τ is an optional time (see Exercise 2), and by assumption it is also predictable. Let $\{\tau_n\}$ be an announcing sequence for τ. By Doob's stopping theorem, for all positive integers n,

$$E[M_\tau \mid \mathcal{F}_{\tau_n}] = M_{\tau_n}.$$

Letting $n \to \infty$ in the above, we obtain

$$E\left[M_\tau \mid \bigvee_{n=1}^{\infty} \mathcal{F}_{\tau_n}\right] = M_{\tau-} \le M_\tau - \varepsilon 1_{\{\tau < \infty\}}.$$

By taking expectations in the above, we obtain: $\varepsilon P(\tau < \infty) \le 0$ and hence $P(\tau < \infty) = 0$. Similarly, for $\sigma \equiv \inf\{t \ge 0 : M_t - M_{t-} \le -\varepsilon\}$ we have $P(\sigma < \infty) = 0$. Since $\varepsilon > 0$ was arbitrary, it follows that P-a.s., M has no jumps at all. ∎

Example. Suppose $\{\mathcal{F}_t\}$ is the filtration generated by a Hunt process (cf. [12, Chapter 3]) with continuous sample paths, where the filtration is augmented by the P-null sets in $\mathcal{F}$. It is known [19] that every (local) martingale adapted to $\{\mathcal{F}_t\}$ has a continuous version. Hence every optional time is predictable. In particular, these properties hold if $\{\mathcal{F}_t\}$ is the standard filtration associated with a d-dimensional Brownian motion.

Next we define a measure on the predictable sets which is the key to the basic isometry used in defining the stochastic integral.

2.4 Measure on the Predictable Sets

Suppose that $Z = \{Z_t,\ t \in \mathbb{R}_+\}$ is a real-valued process adapted to the (standard) filtration $\{\mathcal{F}_t, t \in \mathbb{R}_+\}$, and $Z_t \in L^1$ for each $t \in \mathbb{R}_+$.

We define a set function λ_Z on $\mathcal{R}$ by

$$\lambda_Z\left((s,t] \times F\right) = E\left(1_F(Z_t - Z_s)\right)$$

(2.1)   for $F \in \mathcal{F}_s$ and $s < t$ in $\mathbb{R}_+$,

$$\lambda_Z\left(\{0\} \times F_0\right) = 0 \quad \text{for } F_0 \in \mathcal{F}_0.$$

We extend λ_Z to be a finitely additive set function on the ring $\mathcal{A}$ generated by $\mathcal{R}$ by defining

$$\lambda_Z(A) = \sum_{j=1}^{n} \lambda_Z(R_j)$$

for any $A = \bigcup_{j=1}^{n} R_j$, where $\{R_j, 1 \leq j \leq n\}$ is a finite collection of disjoint sets in $\mathcal{R}$. The value of $\lambda_Z(A)$ is the same for all representations of A as a finite disjoint union of sets in $\mathcal{R}$. We call λ_Z a *content* if $\lambda_Z \geq 0$ on $\mathcal{R}$ and hence on $\mathcal{A}$.

It is clear that if Z is a martingale then $\lambda_Z \equiv 0$, and if Z is a submartingale then $\lambda_Z \geq 0$. In particular, suppose $M = \{M_t, t \in \mathbb{R}_+\}$ is an L^2-martingale, then $(M)^2 = \{(M_t)^2, t \in \mathbb{R}_+\}$ is a submartingale and hence $\lambda_{(M)^2} \geq 0$. More explicitly, for $F \in \mathcal{F}_s$ and $s < t$,

(2.2) $$\lambda_{(M)^2}\left((s,t] \times F\right) = E\left\{1_F(M_t - M_s)^2\right\}.$$

This is proved by setting $Y = 1_F$ in the following important identity. For

$s < t$ in $I\!\!R_+$ and any real-valued $Y \in b\mathcal{F}_s$,

$$
\begin{aligned}
E\left\{Y(M_t - M_s)^2\right\} &= E\left\{Y\left((M_t)^2 - 2M_t M_s + (M_s)^2\right)\right\} \\
&= E\left\{Y\left((M_t)^2 + (M_s)^2\right)\right\} - 2E\left\{YM_s E\left(M_t \mid \mathcal{F}_s\right)\right\} \\
&= E\left\{Y\left((M_t)^2 + (M_s)^2\right)\right\} - 2E\left\{Y(M_s)^2\right\} \\
&= E\left\{Y\left((M_t)^2 - (M_s)^2\right)\right\}.
\end{aligned}
$$

(2.3)

The martingale property of M was used to obtain the third equality above.

We are interested in L^2-martingales M for which $\lambda_{(M)^2}$ can be extended to a measure on $\mathcal{P}$. It is shown in Section 2.8 that if Z is a right continuous positive submartingale, then the content λ_Z can be uniquely extended to a measure on $\mathcal{P}$, and this measure is σ-finite. Setting $Z = M^2$, we see that for a right continuous L^2-martingale M, there is a unique extension of $\lambda_{(M)^2}$ to a (σ-finite) measure on $\mathcal{P}$. An independent proof of this extendibility when M is a *continuous* L^2-martingale is given in Section 4.4.

Until stated otherwise, we suppose that $M = \{M_t, t \in I\!\!R_+\}$ is a right continuous L^2-martingale. We use μ_M to denote the unique measure on $\mathcal{P}$ which extends $\lambda_{(M)^2}$. This measure has been called the Doléans measure of M after C. Doléans-Dade who first made good use of it in a more general setting in [25]. We use $\mathcal{L}^2$ to denote $L^2(I\!\!R_+ \times \Omega, \mathcal{P}, \mu_M)$, unless we need to emphasize the association with M in which case we use $\mathcal{L}^2(\mu_M)$.

Example. Consider a Brownian motion B in $I\!\!R$ with $B_0 \in L^2$ and let $\{\mathcal{F}_t\}$ denote its associated standard filtration. Then $\{B_t, \mathcal{F}_t, t \in I\!\!R_+\}$ is a continuous L^2-martingale. The following calculation shows that μ_B is the product measure $\lambda \times P$ on $\mathcal{P}$, where λ is the Lebesgue measure on $I\!\!R_+$. For $s < t$ and $F \in \mathcal{F}_s$ we have

$$
\begin{aligned}
\lambda_{(B)^2}\left((s,t] \times F\right) &= E\left(1_F (B_t - B_s)^2\right) \\
&= E\left\{1_F E\left((B_t - B_s)^2 \mid \mathcal{F}_s\right)\right\} \\
&= E\left\{(B_t - B_s)^2\right\} E\{1_F\} \\
&= (t - s)P(F) \\
&= (\lambda \times P)\left((s,t] \times F\right)
\end{aligned}
$$

The third equality above follows because $B_t - B_s$ is independent of $\mathcal{F}_s$, a consequence of the independence of the increments of B. The fourth equality follows because $B_t - B_s$ has mean zero and variance $t - s$. For $F_0 \in \mathcal{F}_0$,

$$\lambda_{(B)^2}(\{0\} \times F_0) = 0 = (\lambda \times P)(\{0\} \times F_0).$$

Thus, $\lambda_{(B)^2}$ agrees with $\lambda \times P$ on $\mathcal{R}$ and hence on $\mathcal{A}$. Since $\lambda \times P$ is a measure on $\mathcal{B} \times \mathcal{F} \supset \mathcal{P}$, we have $\mu_B = \lambda \times P$ on $\mathcal{P}$, by the uniqueness of the extension of $\lambda_{(B)^2}$ on $\mathcal{A}$ to μ_B on $\mathcal{P}$.

Example. Consider a Poisson process N with parameter $\alpha > 0$ and let $\{\mathcal{F}_t\}$ denote its associated standard filtration. Then $M \equiv \{N_t - \alpha t, \mathcal{F}_t, t \in \mathbb{R}_+\}$ is a right continuous L^2-martingale. In Exercise 4 you are asked to prove that $\alpha(\lambda \times P)$ is the Doléans measure for M. We shall not consider the Poisson process in detail in this text because stochastic integrals with respect to M can be defined using ordinary Lebesgue-Stieltjes integration (see Exercise 11). In addition, in our subsequent development of the stochastic calculus, from Chapter 4 onwards, we shall restrict ourselves to integrators that are *continuous* local martingales. By restricting to continuous integrators in this way, we are able to present the basic change of variable formula and ideas of stochastic calculus without the cumbersome notation and more elaborate considerations needed when one allows integrators with jumps.

2.5 Definition of the Stochastic Integral

First we define the stochastic integral $\int X \, dM$ when X is an $\mathcal{R}$-simple process and show that the map $X \to \int X \, dM$ is an isometry from a subspace of $\mathcal{L}^2$ into L^2. This isometry is the key to the extension of the definition to all X in $\mathcal{L}^2$.

When X is the indicator function of a predictable rectangle, the integral $\int X \, dM$ is defined as follows. For $s < t$ in $\mathbb{R}_+$ and $F \in \mathcal{F}_s$,

$$(2.4) \qquad \int 1_{(s,t] \times F} \, dM \equiv 1_F(M_t - M_s)$$

and for $F_0 \in \mathcal{F}_0$,

$$(2.5) \qquad \int 1_{\{0\} \times F_0} \, dM \equiv 0.$$

Let $\mathcal{E}$ denote the class of all functions $X : \mathbb{R}_+ \times \Omega \to \mathbb{R}$ that are finite linear combinations of indicator functions of predictable rectangles. Such a function will be called an $\mathcal{R}$-simple process. Thus, $X \in \mathcal{E}$ can be expressed in the form

$$
(2.6) \qquad X = \sum_{j=1}^{n} c_j 1_{(s_j, t_j] \times F_j} + \sum_{k=1}^{m} d_k 1_{\{0\} \times F_{0k}}
$$

where $c_j \in \mathbb{R}$, $F_j \in \mathcal{F}_{s_j}$, $s_j < t_j$ in $\mathbb{R}_+$ for $1 \le j \le n$, $n \in \mathbb{N}$, and $d_k \in \mathbb{R}$, $F_{0k} \in \mathcal{F}_0$ for $1 \le k \le m$, $m \in \mathbb{N}$. This representation, although not unique, can always be chosen such that the predictable rectangles $(s_j, t_j] \times F_j$ for $1 \le j \le n$ and $\{0\} \times F_{0k}$ for $1 \le k \le m$, are disjoint.

The integral $\int X \, dM$ for $X \in \mathcal{E}$ is defined by linearity. Thus, for X of the form (2.6) we have

$$
(2.7) \qquad \int X \, dM \equiv \sum_{j=1}^{n} c_j 1_{F_j} \left(M_{t_j} - M_{s_j} \right).
$$

It can be easily verified that the value of the integral does not depend on the representation chosen for X.

Since $1_R \in \mathcal{L}^2$ for any predictable rectangle R, it follows that $\mathcal{E}$ is a subspace of $\mathcal{L}^2$; and since $M_t \in L^2$ for each t, $\int X \, dM$ is in L^2 for each $X \in \mathcal{E}$. The following theorem shows that the linear map $X \to \int X \, dM$ is an isometry from $\mathcal{E} \subset \mathcal{L}^2$ onto its image in L^2.

Theorem 2.3. *For $X \in \mathcal{E}$ we have the isometry*

$$
(2.8) \qquad E\left\{ \left(\int X \, dM \right)^2 \right\} = \int_{\mathbb{R}_+ \times \Omega} (X)^2 \, d\mu_M.
$$

Proof. Let $X \in \mathcal{E}$ be expressed in the form (2.6) where the predictable rectangles $R_j \equiv (s_j, t_j] \times F_j$ for $1 \le j \le n$ and $\{0\} \times F_{0k}$ for $1 \le k \le m$ are disjoint. Then by (2.7) we have

(2.9)
$$\left(\int X \, dM \right)^2 = \sum_{j=1}^{n} c_j^2 1_{F_j} \left(M_{t_j} - M_{s_j} \right)^2$$
$$+ 2 \sum_{j=1}^{n} \sum_{k=j+1}^{n} c_j c_k 1_{F_j \cap F_k} \left(M_{t_j} - M_{s_j} \right) \left(M_{t_k} - M_{s_k} \right).$$

For $1 \leq j < k \leq n$, since $R_j \cap R_k = \emptyset$, either

(i) $F_j \cap F_k = \emptyset$, or

(ii) $(s_j, t_j] \cap (s_k, t_k] = \emptyset$.

If (i) holds, the term indexed by j and k in the double sum above is zero. If (ii) holds, we may assume without loss of generality that $t_j \leq s_k$. By the martingale property we have $E \left(M_{t_k} - M_{s_k} \,|\, \mathcal{F}_{s_k} \right) = 0$. This implies the basic "orthogonality property" that in the Hilbert space L^2, the increment $M_{t_k} - M_{s_k}$ of M is orthogonal to the subspace $L^2 \left(\Omega, \mathcal{F}_{s_k}, P \right)$, i.e., for any $Y \in L^2 \left(\Omega, \mathcal{F}_{s_k}, P \right)$,

$$E \left\{ Y \left(M_{t_k} - M_{s_k} \right) \right\} = E \left\{ Y E \left(M_{t_k} - M_{s_k} \,|\, \mathcal{F}_{s_k} \right) \right\} = 0.$$

Since $1_{F_j \cap F_k} \left(M_{t_j} - M_{s_j} \right) \in L^2 \left(\Omega, \mathcal{F}_{s_k}, P \right)$, it follows that the expected value of the term indexed by j and k in the double sum in (2.9) is also zero if (ii) holds. Thus, by taking expectations in (2.9) and using (2.1)–(2.2), we obtain

$$E \left\{ \left(\int X \, dM \right)^2 \right\} = \sum_{j=1}^{n} c_j^2 E \left\{ 1_{F_j} \left(M_{t_j} - M_{s_j} \right)^2 \right\}$$
$$= \sum_{j=1}^{n} c_j^2 \mu_M \left((s_j, t_j] \times F_j \right) + \sum_{k=1}^{m} d_k^2 \mu_M \left(\{0\} \times F_{0k} \right)$$
$$= \int_{\mathbb{R}_+ \times \Omega} (X)^2 \, d\mu_M. \quad \blacksquare$$

The extension of the definition of $\int X \, dM$ from integrands X in $\mathcal{E}$ to those in $\mathcal{L}^2$ is based on the isometry (2.8) and the fact that $\mathcal{E}$ is dense in the Hilbert space $\mathcal{L}^2$. A proof of the latter statement is given below.

Lemma 2.4. *The set of $\mathcal{R}$-simple processes $\mathcal{E}$ is dense in the Hilbert space $\mathcal{L}^2$.*

Proof. Since $\mathcal{P}$ is generated by the ring $\mathcal{A}$ and μ_M is σ-finite, then for each $\varepsilon > 0$, and $A \in \mathcal{P}$ such that $\mu_M(A) < \infty$, there is $A_1 \in \mathcal{A}$ such that $\mu_M(A \triangle A_1) < \varepsilon$ where $A \triangle A_1$ is the symmetric difference of A and A_1 (see Halmos [37; p. 42, 49]). It follows that any $\mathcal{P}$-simple function in $\mathcal{L}^2$ can be approximated arbitrarily closely in the $\mathcal{L}^2$-norm by functions in $\mathcal{E}$. The proof is completed by invoking the standard result that the set of $\mathcal{P}$-simple functions is dense in $\mathcal{L}^2$. ■

If we regard $\mathcal{L}^2$ and L^2 as Hilbert spaces, then the map $X \to \int X \, dM$ is a linear isometry from the dense subspace $\mathcal{E}$ of $\mathcal{L}^2$ into L^2, and hence can be uniquely extended to a linear isometry from $\mathcal{L}^2$ into L^2 (see Taylor [74, p. 99]). For $X \in \mathcal{L}^2$, we define $\int X \, dM$ as the image of X under this isometry. Then (2.8) holds for all X in $\mathcal{L}^2$ and we refer to it simply as "the isometry" since it is the only one we use.

Notation. Let $\Lambda^2(\mathcal{P}, M)$ denote the space of all $X \in \mathcal{P}$ such that $1_{[0,t]}X \in \mathcal{L}^2$ for each $t \in I\!\!R_+$. Here $1_{[0,t]}X$ denotes the process defined by

$$(1_{[0,t]}X)(s,\omega) = 1_{[0,t]}(s)X(s,\omega) \quad \text{for all } (s,\omega) \in I\!\!R_+ \times \Omega.$$

Let $X \in \Lambda^2(\mathcal{P}, M)$. For each t, $\int 1_{[0,t]}X \, dM$ is well-defined and has the isometry property:

$$(2.10) \qquad E\left\{ \left(\int 1_{[0,t]}X \, dM \right)^2 \right\} = \int_{[0,t] \times \Omega} (X)^2 \, d\mu_M.$$

By definition, $\mu_M(\{0\} \times \Omega) = 0$, hence by (2.10) we have

$$(2.11) \qquad \int 1_{\{0\} \times \Omega}X \, dM = 0 \quad \text{a.s.}$$

If $X \in \mathcal{E}$ and (2.6) is a representation for X, then for each t, $1_{[0,t]}X$ is in $\mathcal{E}$ and

(2.12) $$\int 1_{[0,t]} X \, dM = \sum_{j=1}^{n} c_j 1_{F_j} \left(M_{t_j \wedge t} - M_{s_j \wedge t} \right).$$

Here the right member of (2.12) is a right continuous L^2-martingale indexed by t. By using the isometry, we shall extend this to prove for $X \in \Lambda^2(\mathcal{P}, M)$ that $\{\int 1_{[0,t]} X \, dM, t \in \mathbb{R}_+\}$ is an L^2-martingale which has a right continuous version; thus showing that these properties of M are preserved by the integration.

Theorem 2.5. *Let $X \in \Lambda^2(\mathcal{P}, M)$ and for each t let $Y_t = \int 1_{[0,t]} X \, dM$. Then $Y = \{Y_t, t \in \mathbb{R}_+\}$ is a zero-mean L^2-martingale and there is a version of Y with all paths right continuous.*

Proof. Let $n \in \mathbb{N}$. Then $1_{[0,n]} X \in \mathcal{L}^2$ and by Lemma 2.4 there is a sequence $\{X^k, k \in \mathbb{N}\}$ in $\mathcal{E}$ which converges to $1_{[0,n]} X$ in $\mathcal{L}^2$. It follows that for each $t \in [0, n]$, $1_{[0,t]} X^k$ converges to $1_{[0,t]} X$ in $\mathcal{L}^2$ as $k \to \infty$, and hence by the isometry, $Y_t^k \equiv \int 1_{[0,t]} X^k \, dM$ converges to $Y_t = \int 1_{[0,t]} X \, dM$ in L^2. For each k, by the remarks following equation (2.12), $Y^k = \{Y_t^k, t \in \mathbb{R}_+\}$ is a right continuous L^2-martingale. Since the martingale property is preserved by L^2-limits (see Proposition 1.3), it follows that $\{Y_t, t \in [0, n]\}$ is an L^2-martingale. Since n was arbitrary, we conclude that $\{Y_t, t \in \mathbb{R}_+\}$ is an L^2-martingale. By (2.11), $Y_0 = 0$ a.s. and hence $E(Y_t) = E(Y_0) = 0$ for all t.

Since $\{Y_t, \mathcal{F}_t, t \in \mathbb{R}_+\}$ is a martingale and $\{\mathcal{F}_t\}$ is a standard filtration, by [12, p. 29], there is a version of $\{Y_t, t \in \mathbb{R}_+\}$ with all paths right continuous. Another proof of this last property of Y can be obtained by replacing "continuous" with "right continuous" in the proof of Theorem 2.6 below. ∎

Theorem 2.6. *Suppose the hypotheses of Theorem 2.5 hold and M has continuous paths. Then there is a version of Y with continuous paths.*

Proof. We first show that for each $n \in \mathbb{N}$ there is a continuous version Z^n of $\{Y_t, t \in [0, n]\}$. For $j < k$ and Y^j, Y^k as in the above proof, $Y^k - Y^j$ is a continuous L^2-martingale and thus by the basic inequality (1.3) of Theorem 1.4 we have

$$(2.13) \qquad P\left(\sup_{0 \leq t \leq n} \left|Y_t^k - Y_t^j\right| \geq \frac{1}{2^m}\right) \leq 2^{2m} E\left(\left|Y_n^k - Y_n^j\right|^2\right)$$

for each $m \in I\!N$. Since Y_n^k converges to Y_n in L^2 as $k \rightarrow \infty$, there is a subsequence $\{Y_n^{k_m}, m \in I\!N\}$ such that

$$(2.14) \qquad E\left(\left|Y_n^{k_{m+1}} - Y_n^{k_m}\right|^2\right) \leq \frac{1}{2^{3m}}.$$

By combining (2.13) and (2.14), we obtain

$$\sum_m P\left(\sup_{0 \leq t \leq n} \left|Y_t^{k_{m+1}} - Y_t^{k_m}\right| \geq \frac{1}{2^m}\right) \leq \sum_m \frac{1}{2^m} < \infty.$$

An application of the Borel-Cantelli lemma then yields

$$P\left(\sup_{0 \leq t \leq n} \left|Y_t^{k_{m+1}} - Y_t^{k_m}\right| \geq \frac{1}{2^m} \quad \text{i.o.}\right) = 0,$$

where i.o. is our abbreviation for "infinitely often". It follows that there is a set Ω_n of probability one such that for each $\omega \in \Omega_n$, $\{Y^{k_m}(t, \omega), m \in I\!N\}$ converges uniformly for $t \in [0, n]$ to some limit $Z^n(t, \omega)$. Since $Y^{k_m}(\cdot, \omega)$ is continuous on $[0, n]$, so is $Z^n(\cdot, \omega)$, by the uniformity of the convergence. Moreover, for each $t \in [0, n]$, $Y_t^{k_m}$ converges a.s. to Z_t^n, and in L^2 to Y_t, as $m \rightarrow \infty$; hence $Z_t^n = Y_t$ a.s. Thus, $Z^n = \{Z_t^n, t \in [0, n]\}$ is a continuous version of $\{Y_t, t \in [0, n]\}$ on Ω_n. For $n_1 < n_2$, $\{Z_t^{n_1}, t \in [0, n_1]\}$ and $\{Z_t^{n_2}, t \in [0, n_1]\}$ are both continuous versions of $\{Y_t, t \in [0, n_1]\}$ on $\Omega_{n_1} \cap \Omega_{n_2}$, and are therefore indistinguishable there. It follows that there is a set $\Omega_0 \subset \bigcap_n \Omega_n$ of probability one such that for each $\omega \in \Omega_0$, $\lim_{n \rightarrow \infty} Z^n(t, \omega)$ exists and is finite for each $t \in I\!R_+$, and for each $n \in I\!N$ this limit equals $Z^n(t, \omega)$ for each $t \in [0, n]$. If we denote this limit by $Z(t, \omega)$, then Z is a continuous version of Y on Ω_0. It can easily be extended to a continuous version on Ω. ∎

Notation. We shall use the notation $\{\int_{[0,t]} X \, dM, t \in I\!R_+\}$ to denote a right continuous version of $\{\int 1_{[0,t]} X \, dM, t \in I\!R_+\}$ and $\int_{(s,t]} X \, dM$ to denote $\int_{[0,t]} X \, dM - \int_{[0,s]} X \, dM$ for $s < t$ in $I\!R_+$. If M is known to be continuous, we shall use $\{\int_0^t X\,dM, t \in \mathbb{R}_+\}$ to denote a continuous version of $\{\int 1_{[0,t]} X\,dM, t \in \mathbb{R}_+\}$ and $\int_s^t X\,dM$ to denote $\int_0^t X\,dM - \int_0^s X\,dM$ for $s < t$.

In the following theorem, we list some properties of the stochastic integral $\int_{[0,t]} X\,dM$.

Theorem 2.7. *Let $X \in \Lambda^2(\mathcal{P}, M)$ and let Y denote the right continuous stochastic integral process $\{\int_{[0,t]} X\,dM, t \in \mathbb{R}_+\}$. Then the following properties hold.*

(i)   *For $s < t$ in $\mathbb{R}_+$ and any r.v. $Z \in b\mathcal{F}_s$, we have $1_{(s,t]}Z \in \mathcal{P}$, $1_{(s,t]}ZX \in \Lambda^2(\mathcal{P}, M)$, and a.s.*

$$(2.15) \qquad \int 1_{(s,t]}ZX\,dM = Z \int_{(s,t]} X\,dM.$$

(ii)   *The measure μ_Y associated with the right continuous L^2-martingale Y has density $(X)^2$ with respect to μ_M, i.e., for any $A \in \mathcal{P}$,*

$$(2.16) \qquad \mu_Y(A) = \int_A (X)^2\,d\mu_M.$$

(iii)   *For any bounded optional time τ,*

$$(2.17) \qquad Y_\tau \equiv \int_{[0,\tau]} X\,dM = \int 1_{[0,\tau]} X\,dM \quad \text{a.s.}$$

Remark. The first equality in (2.17) is by definition, where for each ω, $Y_\tau(\omega)$ is the value of $Y_t(\omega)$ at $t = \tau(\omega)$; whereas the integral on the far right of (2.17) is a random variable defined via the L^2-isometry. Their a.s. equality must therefore be proved.

Proof. For $s < t$ in $\mathbb{R}_+$ and $Z \in \mathcal{F}_s$, $1_{(s,t]}Z \in \mathcal{P}$ follows by linearity and a monotone class argument from the fact that $1_{(s,t] \times G} \in \mathcal{P}$ for $G \in \mathcal{F}_s$.

Then, since $X \in \mathcal{P}$, $1_{(s,t]}ZX \in \mathcal{P}$. (For a partial converse see Exercise 8.) Furthermore, if Z is bounded, then since $X \in \Lambda^2(\mathcal{P}, M)$, we have $1_{(s,t]}ZX \in \Lambda^2(\mathcal{P}, M)$. Now that the measurability and integrability properties in part (i) have been established, we focus on the proof of (2.15). Note that (2.15) is easily verified if $Z = 1_G$ for some $G \in \mathcal{F}_s$ and $X = 1_{(u,v] \times F}$ for some $u < v$ in $\mathbb{R}_+$ and $F \in \mathcal{F}_u$. It then follows by linearity that (2.15) holds when Z is an $\mathcal{F}_s$-simple function and X is in $\mathcal{E}$. For general Z and X, there is a bounded sequence $\{Z^k\}$ of $\mathcal{F}_s$-simple functions converging to Z pointwise on Ω, and a sequence $\{X^k\}$ of functions in $\mathcal{E}$ such that $\lim_{k \to \infty} 1_{(s,t]}X^k = 1_{(s,t]}X$ in $\mathcal{L}^2$. Since $\{Z^k\}$ is bounded, it follows that $\lim_{k \to \infty} 1_{(s,t]}Z^k X^k = 1_{(s,t]}ZX$ in $\mathcal{L}^2$ also. Now,

(2.18)
$$\int 1_{(s,t]}ZX \, dM - Z \int_{(s,t]} X \, dM$$
$$= \int 1_{(s,t]}\left(ZX - Z^k X^k\right) dM$$
$$+ \left\{ \int 1_{(s,t]}Z^k X^k \, dM - Z^k \int_{(s,t]} X^k \, dM \right\}$$
$$+ Z^k \int_{(s,t]} (X^k - X) \, dM + (Z^k - Z) \int_{(s,t]} X \, dM.$$

We claim that the terms following the equals sign above converge to zero in L^1 as $k \to \infty$. By the simple function case discussed above, the second term (in braces) is zero. The first and third terms converge to zero in L^2, by the isometry. The last term tends to zero in L^1, by Schwarz's inequality and bounded convergence. Since the expression in (2.18) preceding the equals sign is independent of k, it follows that it is zero a.s., proving (i).

For the proof of part (ii), it suffices to prove (2.16) for $A \in \mathcal{R}$, since the measures μ_Y and $(X)^2 d\mu_M$ on $\mathcal{P}$ are uniquely determined by their values on $\mathcal{R}$. If $A = \{0\} \times F_0$ for $F_0 \in \mathcal{F}_0$, both sides of (2.16) are zero. On the other hand, if $A = 1_{(s,t] \times F}$ for some $s < t$ and $F \in \mathcal{F}_s$, then

$$\mu_Y(A) = E\left\{1_F(Y_t - Y_s)^2\right\} = E\left\{(1_F \int_{(s,t]} X \, dM)^2\right\}$$

which by part (i) equals

$$E\left\{\left(\int 1_{(s,t]\times F} X\, dM\right)^2\right\} = \int 1_{(s,t]\times F}(X)^2\, d\mu_M = \int_A (X)^2\, d\mu_M.$$

The first equality above follows by the isometry. Thus (2.16) holds for all A in $\mathcal{R}$ and hence for all A in $\mathcal{P}$.

For the proof of part (iii), let τ be an optional time, bounded by C, say. We approximate τ in the standard way by a sequence $\{\tau_n,\, n \in I\!N\}$ of optional times such that for each n, τ_n takes only finitely many values and (2.17) holds with τ_n in place of τ.

As in the proof of Lemma 2.1, for each n let $\tau_n = 2^{-n}\left[2^n\tau + 1\right]$. Also let $N_n = [2^n C]$. Then

$$(2.19) \qquad [0,\tau_n] = (\{0\} \times \Omega) \cup \bigcup_{k=0}^{N_n} (k2^{-n},(k+1)2^{-n}] \times \{\tau \geq k2^{-n}\}$$

is in $\mathcal{A}$ and by the boundedness of τ_n, $1_{[0,\tau_n]}X \in \mathcal{L}^2$. Now, for each n,

$$Y_{\tau_n} = \sum_{k=0}^{N_n} 1_{\{k2^{-n}\leq\tau<(k+1)2^{-n}\}} Y_{(k+1)2^{-n}}$$

$$= \sum_{k=0}^{N_n} 1_{\{\tau\geq k2^{-n}\}} \left(Y_{(k+1)2^{-n}} - Y_{k2^{-n}}\right).$$

Here the second equality is obtained by partial summation using $Y_0 = 0$ and $0 \leq \tau < (N_n + 1)2^{-n}$. Thus by the definition of Y_t and part (i) we have a.s.

$$Y_{\tau_n} = \sum_{k=0}^{N_n} \int 1_{(k2^{-n},(k+1)2^{-n}]\times\{\tau\geq k2^{-n}\}} X\, dM.$$

By linearity, (2.11), and (2.19), it follows that a.s.

$$(2.20) \qquad\qquad Y_{\tau_n} = \int 1_{[0,\tau_n]} X\, dM.$$

Since $\tau_n \downarrow \tau$ and Y is right continuous, the left side of (2.20) converges pointwise on Ω to Y_τ as $n \to \infty$; and since τ_n is bounded by $C+1$, it follows by dominated convergence and the isometry that the right side converges to $\int 1_{[0,\tau]} X \, dM$ in L^2. Hence (2.17) holds. ∎

The following corollary will be needed in the next section.

Corollary 2.8. *Let $s < t$ in $\mathbb{R}_+$, $F \in \mathcal{F}_s$, and τ be an optional time. Then we have a.s.:*

$$(2.21) \qquad \int 1_{[0,\tau]} 1_{(s,t] \times F} \, dM = 1_F (M_{t \wedge \tau} - M_{s \wedge \tau}).$$

Proof. Let $X = 1_{(s,t] \times F}$. Then,

$$\int 1_{[0,u]} X \, dM = 1_F (M_{t \wedge u} - M_{s \wedge u}).$$

The right side of the above equality is right continuous in u and therefore may be used as the right continuous version $\int_{[0,u]} X \, dM$ of the left side. By replacing u by $\tau \wedge t$, we obtain

$$\int_{[0,\tau \wedge t]} X \, dM = 1_F (M_{t \wedge \tau} - M_{s \wedge \tau}).$$

It follows from (2.17) with $\tau \wedge t$ in place of τ there, that the left side of the above is equal a.s. to the left side of (2.21). Hence (2.21) holds a.s. ∎

Our definitions of the measure μ_M and the stochastic integral $\int_{[0,t]} X \, dM$ only involved the increments of M. Hence the values of these quantities would remain unchanged if we replaced M by $M - M_0$ in their definitions. Indeed, the following depends on this.

2.6 Extension to Local Integrators and Integrands

So far we have considered stochastic integrals $\int_{[0,t]} X \, dM$ where the integrator is a right continuous L^2-martingale and the integrand is in $\Lambda^2(\mathcal{P}, M)$. As a final extension we shall define the stochastic integral for integrators and integrands which only possess these properties in a local sense. Consequently, we shall no longer assume that M is a right continuous L^2-martingale. Instead, for the rest of this chapter, we suppose that M is a right continuous *local* L^2-martingale (see Section 1.10 for the definition). If $\{\tau_k\}$ is a localizing sequence for M, we use M^k to denote the right continuous L^2-martingale $\{M_{t \wedge \tau_k} - M_0, t \in \mathbb{R}_+\}$ for each k.

Next we define the class of integrands associated with M.

Definition. Let $\Lambda(\mathcal{P}, M)$ denote the class of all processes X for which there is a localizing sequence $\{\tau_k\}$ for M such that M^k is an L^2-martingale and

(2.22) $1_{[0,\tau_k]} X \in \Lambda^2(\mathcal{P}, M^k)$ for each k.

Such a sequence will be called a localizing sequence for (X, M).

Example. Suppose M has continuous paths and X is a continuous adapted process. We claim $X \in \Lambda(\mathcal{P}, M)$ and

$$\tau_k = \inf\{t > 0 : |M_t - M_0| \vee |X_t| > k\}$$

defines a localizing sequence for (X, M).

For the proof of this claim we note that by results in Chapter 3, X is predictable. By the definition of τ_k, $(X)^2 \leq k^2$ on $(0, t \wedge \tau_k]$. Moreover, by the isometry and Theorem 2.7(iii) we have

$$\int_{\mathbb{R}_+ \times \Omega} 1_{(0, t \wedge \tau_k]}\, d\mu_{M^k} = E\left\{(M_{t \wedge \tau_k} - M_0)^2\right\} \leq k^2.$$

Thus, by combining the above with the fact that $\mu_{M^k}(\{0\} \times \Omega) = 0$, we obtain

$$\int_{\mathbb{R}_+ \times \Omega} 1_{[0, t \wedge \tau_k]}(X)^2\, d\mu_{M^k} \leq k^4,$$

which proves the assertion. ∎

An important special case of the above example is obtained by setting $X = M$.

Let $X \in \Lambda(\mathcal{P}, M)$ and $\{\tau_k\}$ be a localizing sequence for (X, M). Then $Y^k \equiv \{\int_{[0,t]} 1_{[0,\tau_k]} X\, dM^k, t \in \mathbb{R}_+\}$ is a right continuous L^2-martingale for each k, by the notational convention following Theorem 2.6. We shall define $Y = \{\int_{[0,t]} X\,dM, t \in \mathbb{R}_+\}$ as the a.s. limit of the Y^k's, just as Z was defined from the Z^n's in the proof of Theorem 2.6. The difference being that here we use random truncation times τ_k whereas constant times n were used before. To validate this procedure, we need to verify that the following consistency condition holds:

(i)   for each k, for almost every ω:

(2.23) $Y_t^m(\omega) = Y_t^k(\omega)$ for all $t \in [0, \tau_k]$ and $m \geq k$,

and to show that

(ii)   the definition of Y is independent (up to indistinguishability) of the choice of a localizing sequence for (X, M).

These assertions are formally obvious, but their proofs are long in details. They follow from the two lemmas below which are spelled out for the meticulous reader.

Lemma 2.9. *Let τ and η be optional times such that $M^\tau = \{M_{t \wedge \tau} - M_0, t \in \mathbb{R}_+\}$ and $M^\eta = \{M_{t \wedge \eta} - M_0, t \in \mathbb{R}_+\}$ are right continuous L^2-martingales. Let μ^τ and μ^η denote the measures μ_{M^τ} and μ_{M^η} on $\mathcal{P}$ associated respectively with M^τ and M^η. Then μ^τ and μ^η induce the same measure on the stochastic interval $[0, \tau \wedge \eta]$, i.e., for each $A \in \mathcal{P}$:*

(2.24) $\mu^\tau(A \cap [0, \tau \wedge \eta]) = \mu^\eta(A \cap [0, \tau \wedge \eta]).$

Proof. Since the predictable rectangles generate $\mathcal{P}$, it suffices to prove (2.24) when A is a predictable rectangle. Clearly both sides of (2.24) are zero when $A = \{0\} \times F_0$ for some $F_0 \in \mathcal{F}_0$. On the other hand, if $A = (s, t] \times F$ for some $s < t$ and $F \in \mathcal{F}_s$, then by the isometry

$$\mu^\tau(A \cap [0, \tau \wedge \eta]) = E\left\{ \left(\int 1_{[0, \tau \wedge \eta]} 1_{(s,t] \times F}\, dM^\tau \right)^2 \right\}.$$

By Corollary 2.8 and since $M_u^\tau = M_u - M_0$ for $0 \leq u \leq \tau$, the right side above equals

$$E\{1_F(M_{t \wedge \tau \wedge \eta} - M_{s \wedge \tau \wedge \eta})^2\}.$$

Since the last expression is symmetric in τ and η, (2.24) follows for $A = (s,t] \times F$, and hence for all A in $\mathcal{P}$. ∎

Lemma 2.10. *Let τ and η be optional times such that $M^\tau = \{M_{t \wedge \tau} - M_0, t \in \mathbb{R}_+\}$ and $M^\eta = \{M_{t \wedge \eta} - M_0, t \in \mathbb{R}_+\}$ are right continuous L^2-martingales, and $1_{[0,\tau]}X \in \Lambda^2(\mathcal{P}, M^\tau)$ and $1_{[0,\eta]}X \in \Lambda^2(\mathcal{P}, M^\eta)$. Let Y^τ and Y^η respectively denote the right continuous L^2-martingales $\{\int_{[0,t]} 1_{[0,\tau]}X \, dM^\tau, t \in \mathbb{R}_+\}$ and $\{\int_{[0,t]} 1_{[0,\eta]}X \, dM^\eta, t \in \mathbb{R}_+\}$. Then*

$$(2.25) \qquad P\{Y_t^\tau = Y_t^\eta \text{ for } 0 \leq t \leq \tau \wedge \eta\} = 1.$$

Proof. To prove (2.25), it is equivalent to prove the processes $\{Y_{t \wedge \tau \wedge \eta}^\tau, t \geq 0\}$ and $\{Y_{t \wedge \tau \wedge \eta}^\eta, t \geq 0\}$ are indistinguishable, and since these are right continuous, it suffices to prove

$$(2.26) \qquad Y_{t \wedge \tau \wedge \eta}^\tau = Y_{t \wedge \tau \wedge \eta}^\eta \quad \text{a.s.}$$

for each t. It is easily verified using (2.17) and (2.21) that (2.26) holds if X is the indicator function of a predictable rectangle, and hence by linearity if X is in $\mathcal{E}$. For the general case, using the same notation as in Lemma 2.9, we have $1_{[0,t \wedge \tau \wedge \eta]}X \in \mathcal{L}^2(\mu^\tau)$. Hence there is a sequence $\{X^n\}$ in $\mathcal{E}$ which converges to $1_{[0,t \wedge \tau \wedge \eta]}X$ in $\mathcal{L}^2(\mu^\tau)$ and therefore $1_{[0,t \wedge \tau \wedge \eta]}(X^n - X) \to 0$ in $\mathcal{L}^2(\mu^\tau)$ as $n \to \infty$. The latter convergence is also in $\mathcal{L}^2(\mu^\eta)$, since μ^τ and μ^η induce the same measure on $[0, t \wedge \tau \wedge \eta]$ by Lemma 2.9. We have already verified that (2.26) holds if X is replaced by X^n. By letting $n \to \infty$ and using (2.17) and the isometries (for τ and η), it follows that (2.26) holds for X. ∎

By setting $\tau = \tau_m$ and $\eta = \tau_k$ in Lemma 2.10, we obtain (2.23). Consequently, there is a set Ω_0 of probability one such that for each $\omega \in \Omega_0, \lim_{m \to \infty} Y^m(t, \omega)$ exists and is finite for each t, and for each k and $t \in [0, \tau_k]$ this limit equals $Y^k(t, \omega)$. We denote this limit by $Y(t, \omega)$. Then $Y(\cdot, \omega)$ is right continuous for each $\omega \in \Omega_0$ and can easily be defined so that it is right continuous for $\omega \in \Omega \backslash \Omega_0$. Then for each k, almost surely, $Y_{t \wedge \tau_k} = Y_t^k$ for all t. Hence Y is a right continuous local L^2-martingale with localizing sequence $\{\tau_k\}$.

We shall denote Y_t by $\int_{[0,t]} X \, dM$ and $Y_t - Y_s$ by $\int_{(s,t]} X \, dM$. If M is actually continuous, then so is Y, and Y_t will be denoted by $\int_0^t X \, dM$ and $Y_t - Y_s$ by $\int_s^t X \, dM$.

The fact that the definition of Y is independent (up to indistinguishability) of the choice of a localizing sequence for (X, M) is an easy consequence of Lemma 2.10. The formal proof is left to the reader.

The following theorem is an immediate consequence of the above discussion and the example which follows (2.22). Recall that a continuous local martingale is automatically a local L^2-martingale.

Theorem 2.11. *Let M be a continuous local martingale and X be a continuous adapted process. Then $X \in \Lambda(\mathcal{P}, M)$ and $\{\int_0^t X \, dM, t \in \mathbb{R}_+\}$ is a continuous local martingale.*

If M is a right continuous L^2-martingale and $X \in \Lambda^2(\mathcal{P}, M)$, the above definition of $\int_{[0,t]} X \, dM$ is consistent with that given in the previous section because the integrals are unchanged if M is replaced by $M - M_0$. This replacement is also used to simplify some later proofs by reducing to the case $M_0 = 0$ so that $M_t^k = M_{t \wedge \tau_k}$. In connection with this, we emphasize that only the integrator can be replaced by $M - M_0$. In particular if M is a continuous local martingale we have

$$
\begin{aligned}
\int_0^t M \, dM &= \int_0^t (M - M_0) \, dM + M_0(M_t - M_0) \\
&= \int_0^t (M - M_0) \, d(M - M_0) + M_0(M_t - M_0).
\end{aligned}
$$

(2.27)

2.7 Substitution Formula

Theorem 2.12. *Let M be a right continuous local martingale, $X \in \Lambda(\mathcal{P}, M)$ and $Y_t = \int_{[0,t]} X dM$ for all $t \geq 0$. Suppose $Z \in \Lambda(\mathcal{P}, Y)$. Then $XZ \in \Lambda(\mathcal{P}, M)$ and a.s. for all $t \geq 0$,*

$$(2.28) \qquad \int_{[0,t]} Z dY = \int_{[0,t]} XZ dM.$$

Proof. By taking the minimum of a localizing sequence $\{\sigma_n\}$ for (X, M) (see (2.22)) and a localizing sequence $\{\rho_n\}$ for (Z, Y), we obtain a sequence $\{\tau_n \equiv \sigma_n \wedge \rho_n\}$ that is simultaneously localizing for (X, M) and (Z, Y). By stopping M and Y with this sequence and multiplying X and Z by $1_{[0,\tau_n]}$, we see that it suffices to prove the theorem for the case in which M and Y are L^2-martingales, $X \in \Lambda^2(\mathcal{P}, M)$ and $Z \in \Lambda^2(\mathcal{P}, Y)$. The proof for this case is divided into three parts.

(i)   *Suppose Z is an $\mathcal{R}$-simple integrand. Then, for each $t \in \mathbb{R}_+$, (2.28) holds a.s.*

Proof of (i). Note that $1_{[0,t]} XZ \in \mathcal{L}^2(\mu_M)$, since this holds with X in place of XZ and Z is bounded. It follows from (2.5) with Y in place of M, and (2.10) with XZ in place of X and $\mu_M(\{0\} \times \Omega) = 0$, that (2.28) holds when $Z = 1_{\{0\} \times F_0}$ where $F_0 \in \mathcal{F}_0$. By Theorem 2.7, (2.28) also holds when $Z = 1_{(r,s] \times F_r}$ for $F_r \in \mathcal{F}_r$, $0 \leq r < s < \infty$. It then follows by linearity that (2.28) holds for any $\mathcal{R}$-simple Z.

(ii)   *For each $t \in \mathbb{R}_+$,*

$$(2.29) \qquad \int 1_{[0,t]} Z^2 d\mu_Y = \int 1_{[0,t]} X^2 Z^2 d\mu_M,$$

and hence $XZ \in \Lambda^2(\mathcal{P}, M)$.

Proof of (ii). If Z is $\mathcal{R}$-simple, then by the isometry (2.8) and part (i) above we have

$$\int 1_{[0,t]} Z^2 d\mu_Y = E\left[\left(\int_{[0,t]} Z dY\right)^2\right]$$

(2.30)

$$= E\left[\left(\int_{[0,t]} XZ dM\right)^2\right]$$

$$= \int 1_{[0,t]} X^2 Z^2 d\mu_M.$$

The result for $Z \in \Lambda^2(\mathcal{P}, Y)$ then follows by applying a monotone class theorem.

(iii) *Almost surely, (2.28) holds for all $t \in \mathbb{R}_+$.*

Proof of (iii). Since both sides of (2.28) are right continuous processes, it suffices to show that (2.28) holds P-a.s for fixed t. By part (i) above, this holds for all $\mathcal{R}$-simple functions Z. For a general $Z \in \Lambda^2(\mathcal{P}, Y)$, there is a sequence $\{Z^{(m)}\}$ of $\mathcal{R}$-simple integrands such that $1_{[0,t]} Z^{(m)}$ converges to $1_{[0,t]} Z$ in $\mathcal{L}^2(\mu_Y)$ as $m \to \infty$. It follows by the L^2-isometry that

(2.31)
$$\int_{[0,t]} Z^{(m)} dY \to \int_{[0,t]} Z dY \quad \text{in } L^2.$$

Hence, using (2.29) with Z replaced by $Z^{(m)} - Z$, we see that as $m \to \infty$, $1_{[0,t]} X(Z^{(m)} - Z)$ converges to zero in $\mathcal{L}^2(\mu_M)$, and so by the isometry (2.8),

(2.32)
$$\int_{[0,t]} XZ^{(m)} dM \to \int_{[0,t]} XZ dM \quad \text{in } L^2.$$

Since the $Z^{(m)}$'s are $\mathcal{R}$-simple, the left members of (2.31) and (2.32) are equal for each m, by part (i), and hence the right members are equal P-a.s. ∎

2.8 A Sufficient Condition for Extendibility of λ_Z

For the proofs below, we have benefited from the presentation given in Letta [53].

For any set $A \subset \mathbb{R}_+ \times \Omega$ and $\omega \in \Omega$, let A^ω denote the ω-section of A:

$$A^\omega = \{t \in \mathbb{R}_+ : (t, \omega) \in A\}.$$

The *début* D_A of A is defined by

(2.33) $$D_A(\omega) = \inf A^\omega, \quad \text{for all} \quad \omega \in A,$$

where $\inf \emptyset = \infty$.

Lemma 2.13.

(i)   If $B_j \subset \mathbb{R}_+$ for $1 \leq j \leq k \leq \infty$, then

$$\inf_{1 \leq j \leq k} (\inf B_j) = \inf \left(\bigcup_{j=1}^{k} B_j \right).$$

(ii)   If for each $n \geq 1$, C_n is a compact subset of $\mathbb{R}_+$ and $C_n \supset C_{n+1}$ for all n, then

$$\lim_{n \to \infty} \uparrow (\inf C_n) = \inf \left(\bigcap_{n=1}^{\infty} C_n \right).$$

Remark. The empty set $\emptyset$ is a compact set.

Proof. We prove part (ii) only. If there is an n such that $C_n = \emptyset$, then $\inf C_n = \infty$ and the result reduces to $\infty = \infty$. On the other hand, suppose none of the C_n is empty, let $C = \bigcap_{n=1}^{\infty} C_n$ and $t_n = \inf C_n$ for each n. Then, $t_n \in C_n$ and $t^* \equiv \lim_{m \to \infty} \uparrow t_m \in C_n$ for all $n \geq 1$. Thus, $t^* \in C$, $t^* \geq \inf C \geq t_n$ for all n. Hence $t^* = \inf C$.   ∎

Example. If the sets in Lemma 2.13(ii) are not compact, the result can fail, as the following example illustrates. If $C_n = (1 - \frac{1}{n}, 1) \cup \{2\}$, then $C_n \downarrow \{2\}$, but $\inf C_n = 1 - \frac{1}{n} \uparrow 1 \neq \inf \bigcap_n C_n = 2$.

A subset A of $I\!R_+ \times \Omega$ is called a *stochastic compact* if for each $\omega \in \Omega$, A^ω is a compact subset of $I\!R_+$.

Lemma 2.14.

(i)   If $A_j \subset I\!R_+ \times \Omega$ for $1 \leq j \leq k \leq \infty$, then

$$\inf_{1 \leq j \leq k} D_{A_j} = D_{\cup_{j=1}^k A_j}.$$

(ii) If for each $n \geq 1$, A_n is a stochastic compact and $A_n \supset A_{n+1}$ for all n, then

$$\lim_{n \to \infty} \uparrow D_{A_n} = D_{\cap_{n=1}^\infty A_n}.$$

Proof. This is an immediate consequence of Lemma 2.13. ∎

Definition. A random variable X is said to have *finite range* if $\{X(\omega) : \omega \in \Omega\}$ is a finite set.

Lemma 2.15. *If $A \in \mathcal{A}$, then D_A is optional and has finite range.*

Proof. Since $A \in \mathcal{A}$, there are finitely many disjoint predictable rectangles $R_1, \ldots, R_n$, such that $A = \bigcup_{j=1}^n R_j$. By Lemma 2.14(i),

$$D_A = \min_{1 \leq j \leq n} D_{R_j}.$$

If $R = (s,t] \times F_s$ for $F_s \in \mathcal{F}_s$, then $D_R = s 1_{F_s} + \infty \cdot 1_{F_s^c}$ is an optional time. If $R = \{0\} \times F_0$ for $F_0 \in \mathcal{F}_0$, then $D_R = 0 \cdot 1_{F_0} + \infty \cdot 1_{F_0^c}$ is an optional time. Since the minimum of a finite number of optional times, each of which has finite range, is optional and has finite range, the result follows. ∎

Remark. More generally, the début of any optional (in fact, of any progressively measurable) set is optional (see Chung [12, Theorem 3, Section 1.5]).

Theorem 2.16. *Let Z be a right continuous positive submartingale. Then the content λ_Z defined in Section 2.4 can be uniquely extended to a measure on $\mathcal{P}$, and this measure is σ-finite.*

Proof. We first note for later reference that for each $t \geq 0$, from the non-negativity and submartingale property of Z,

$$0 \leq Z_s \leq E[Z_t \,|\, \mathcal{F}_s] \quad \text{for all} \quad 0 \leq s \leq t,$$

and hence $\{Z_s, 0 \leq s \leq t\}$ is uniformly integrable.

Since

$$\lambda_Z([0, t] \times \Omega) = E[Z_t - Z_0] < \infty \quad \text{for all} \quad t \geq 0,$$

any extension of λ_Z to a measure on $\mathcal{P}$ must be σ-finite.

By the Caratheodory extension theorem [37, p. 54], since the σ-ring generated by $\mathcal{A}$ is $\mathcal{P}$, to prove that λ_Z is uniquely extendible to a measure on $\mathcal{P}$, it suffices to prove λ_Z is countably additive on $\mathcal{A}$. For this it is enough to show that for any sequence $\{G_n\} \subset \mathcal{A}$ such that $G_n \downarrow \emptyset$, $\lambda_Z(G_n) \to 0$. Now for any $G \in \mathcal{A}$,

$$\begin{aligned}\lambda_Z(G) &= \lambda_Z\left(G \cap (\{0\} \times \Omega)\right) + \lambda_Z\left(G \cap ((0, \infty) \times \Omega)\right) \\ &= \lambda_Z\left(G \cap ((0, \infty) \times \Omega)\right),\end{aligned}$$

where $G \cap ((0, \infty) \times \Omega) \in \mathcal{A}$. Thus, it suffices to consider $\{G_n\} \subset \mathcal{A} \cap ((0, \infty) \times \Omega)$. Note that if $G = (s, t] \times F_s$ for $F_s \in \mathcal{F}_s$, then for any $m \in I\!\!N$ such that $s + \frac{1}{m} < t$, $H \equiv (s + \frac{1}{m}, t] \times F_s \in \mathcal{R}$, $K \equiv [s + \frac{1}{m}, t] \times F_s$ is a stochastic compact, and $H \subset K \subset G$. Furthermore,

$$\lambda_Z(G \backslash H) = \lambda_Z(G) - \lambda_Z(H) = E\left(1_{F_s}(Z_{s + \frac{1}{m}} - Z_s)\right) \to 0$$

as $m \to \infty$, by the right continuity of Z and the uniform integrability of $\{Z_u : s \leq u \leq s + 1\}$. Since each set in $\mathcal{A} \cap ((0, \infty) \times \Omega)$ is a finite union of disjoint sets in $\mathcal{R} \cap ((0, \infty) \times \Omega)$, it follows that given any $\varepsilon > 0$, for each $j \geq 1$, there is $H_j \in \mathcal{A}$ and a stochastic compact K_j such that

$H_j \subset K_j \subset G_j$ and $\lambda_Z(G_j) - \lambda_Z(H_j) < \varepsilon 2^{-j}$. Since $\{G_n\}$ is decreasing, we have for each n,

$$\hat{H}_n \equiv \bigcap_{j=1}^{n} H_j \subset \hat{K}_n \equiv \bigcap_{j=1}^{n} K_j \subset G_n \equiv \bigcap_{j=1}^{n} G_j,$$

where $\hat{H}_n \in \mathcal{A}$, $\hat{K}_n$ is a stochastic compact, and

(2.34) $\qquad \lambda_Z(G_n \backslash \hat{H}_n) = \lambda_Z(G_n) - \lambda_Z(\hat{H}_n) < \varepsilon.$

By hypothesis $G_n \downarrow \emptyset$, hence $\hat{K}_n \downarrow \emptyset$, and then by Lemma 2.14(ii),

$$D_{\hat{K}_n} \uparrow D_\emptyset = \infty.$$

Since $T_n \equiv D_{\hat{H}_n} \geq D_{\hat{K}_n}$, we have $T_n \uparrow \infty$ as $n \to \infty$. Thus, for any $t \geq 0$ and $\omega \in \Omega$, there is $N(\omega) \in I\!N$ such that $T_n(\omega) \wedge t = t$ for all $n \geq N(\omega)$. Thus, $Z_{T_n \wedge t} \to Z_t$ pointwise on Ω as $n \to \infty$. Now, by Lemma 2.15, T_n is optional and has finite range. Hence $T_n \wedge t$ also has these properties, and then by the submartingale property of Z,

$$0 \leq Z_{T_n \wedge t} \leq E[Z_t \,|\, \mathcal{F}_{T_n \wedge t}] \quad \text{for all } n \geq 1.$$

(Note we did not need to use the right continuity of Z for this.) Consequently, $\{Z_{T_n \wedge t}, n \geq 1\}$ is uniformly integrable and it follows that

$$\lim_{n \to \infty} E[Z_{T_n \wedge t}] = E[Z_t].$$

By the definition of λ_Z,

(2.35) $\qquad \lambda_Z\left((T_n \wedge t, t]\right) = E[Z_t - Z_{T_n \wedge t}] \to 0 \quad \text{as } n \to \infty.$

Since $G_1 \in \mathcal{A} \cap ((0, \infty) \times \Omega)$, there is $t > 0$ such that $G_1 \subset [0, t]$ and so $\hat{H}_n \subset [0, t]$. On the other hand, by the form of the sets in $\mathcal{A} \cap ((0, \infty) \times \Omega)$, the graph $\{(D_{\hat{H}_n}(\omega), \omega) : \omega \in \Omega\} \subset \overline{I\!R}_+ \times \Omega$ of the début of the set $\hat{H}_n$ does not meet $\hat{H}_n$, and so $\hat{H}_n \subset (T_n, \infty)$. Hence, $\hat{H}_n \subset (T_n \wedge t, t]$. It follows from (2.35) that $\lambda_Z(\hat{H}_n) \to 0$ as $n \to \infty$, and hence by (2.34), $\lambda_Z(G_n) \to 0$ as $n \to \infty$, as desired. ∎

Corollary 2.17. *If M is a right continuous L^2-martingale, then $\lambda_{(M)^2}$ has a unique extension to a measure on $\mathcal{P}$, and this measure is σ-finite.*

2.9 Exercises

1. Let $\mathcal{G}_1$ denote the σ-field generated by all stochastic intervals of the form $[\eta, \tau]$ where η and τ are optional times. Similarly, let $\mathcal{G}_2$, $\mathcal{G}_3$ and $\mathcal{G}_4$ denote the σ-fields generated by the stochastic intervals of the form $[\eta, \tau)$, $(\eta, \tau]$ and (η, τ), respectively. Can you identify $\mathcal{G}_i$ ($i \in \{1, 2, 3, 4\}$) with $\mathcal{P}$, $\mathcal{O}$ or neither?
(*Answer:* $\mathcal{G}_1 = \mathcal{G}_2 = \mathcal{O}$, $\mathcal{G}_3 \subset \mathcal{P}$ but $\mathcal{G}_3 \neq \mathcal{P}$, and $\mathcal{G}_4 \subset \mathcal{O}$ but $\mathcal{G}_4 \neq \mathcal{O}$.)

2. Suppose $X = \{X_t, t \geq 0\}$ is a right continuous process that has finite left limits on $(0, \infty)$ and is adapted to $\{\mathcal{F}_t, t \geq 0\}$. Prove that $\tau \equiv \inf\{t \geq 0 : X_t - X_{t-} \geq \varepsilon\}$ is an optional time.
(This result can be proved from first principles, but it is also a special case of the optionality of the début of a progressively measurable set, cf. Chung [12, Section 1.5].)

In the next two exercises, let $\{N_t, t \in \mathbb{R}_+\}$ be a Poisson process with parameter $\alpha > 0$ and let $\{\mathcal{F}_t, t \geq 0\}$ be the associated standard filtration as defined in Section 1.8.

3. Let T be the time of the first jump of N. Show the following.
(a) T is an optional time.
(b) T is *not* a predictable time.
Hint for (b): For a proof by contradiction, suppose that $\{T_n\}$ is an announcing sequence for T. Then use the memoryless property of the Poisson process at the optional times T_n to conclude that for any $\varepsilon > 0$:

$$P(T - T_n > \varepsilon) = P(T > \varepsilon).$$

Since $T_n \uparrow T$ P-a.s., the left member above tends to zero as $n \to \infty$. On the other hand, the right member is independent of n and strictly positive.

4. Show that $M_t = N_t - \alpha t$ defines a (right continuous) L^2-martingale and that its Doléans measure is $\mu_M = \alpha(\lambda \times P)$ (cf. Exercise 2 of Chapter 1).

5. Let T be an optional time relative to $\{\mathcal{F}_t\}$. Suppose the following two properties hold:

(i) for each $t > 0$, $P(0 < T \leq t) > 0$, and
(ii) for $A \in \mathcal{F}_t$, $P(A \cap \{T > t\})$ is either equal to 0 or to $P(T > t)$.

Prove that if S is also an optional time relative to $\{\mathcal{F}_t\}$, then

$$P(0 < S < T) < P(T > 0).$$

This implies that T is not predictable. Since the first jump time of a Poisson process satisfies (i)-(ii), this gives an alternative proof of 3(b) above.

6. Let $(W, \mathcal{G}, \nu)$ be a σ-finite measure space where the σ-field $\mathcal{G}$ is generated by a ring $\mathcal{H}$ such that $\nu(A) < \infty$ for each $A \in \mathcal{H}$. Prove that if $A_1 \in \mathcal{G}$ and $\nu(A_1) < \infty$, then for each $\varepsilon > 0$ there is $A_2 \in \mathcal{H}$ such that $\nu(A_1 \triangle A_2) < \varepsilon$.

7. Use the monotone class theorem for functions (see Section 1.2) to give an alternative proof of Lemma 2.4.
Hint: First prove this result for bounded functions in $L^2([0, t] \times \Omega, \mathcal{P}, \mu_M)$.

8. Suppose X is a predictable process and $X > 0$. Prove that for any $0 \leq s < t < \infty$ and random variable $Z \in \mathcal{F}$, $1_{(s,t]} Z X$ is predictable if and only if Z is $\mathcal{F}_s$-measurable.

9. Prove that (2.17) is also true if τ is a finite optional time and $X \in \mathcal{L}^2$.

10. Let $\{\tau_k\}$ be a localizing sequence for a right continuous local L^2-martingale M. For fixed k, define

$$Z_t = M_{t \wedge \tau_k \wedge k} - M_0 \quad \text{for all} \quad t \geq 0.$$

Show that Z is an L^2-martingale and that its Doléans measure μ_Z is finite on $(\mathbb{R}_+ \times \Omega, \mathcal{P})$.

11. Let N be a Poisson process with parameter $\alpha > 0$. Define $M_t = N_t - \alpha t$. Note from Exercise 4 that $\mu_M = \alpha(\lambda \times P)$. Suppose $X \in \Lambda^2(\mathcal{P}, M)$. Then the stochastic integral $\int_{[0,t]} X \, dM$ is well defined for all $t \in \mathbb{R}_+$. Prove that for each t, $\int_{[0,t]} X_s \, dN_s$ and $\int_{[0,t]} X_s \, ds$ are almost surely well defined as Lebesgue-Stieltjes integrals and that they define random variables satisfying

$$(2.36) \qquad \int_{[0,t]} X_s \, dN_s = \int_{[0,t]} X \, dM + \alpha \int_{[0,t]} X_s \, ds,$$

where the integral with respect to dM is a stochastic integral defined by the L^2-isometry.

Hint: First prove (2.36) for an $\mathcal{R}$-simple X and then use a monotone class argument to extend to $X \in \Lambda^2(\mathcal{P}, M)$.

In Exercises 12 and 13 below, B is a Brownian motion in $\mathbb{R}$ with $B_0 \in L^2$. You may use the result from the first example in Section 4.2 in solving these exercises.

12. Working from the fundamental definition, evaluate the stochastic integral $\int_0^t B_s \, dB_s$. If you are ambitious, try also $\int_0^t B_s^k \, dB_s$ for $k = 2, 3, \ldots$, where B_s^k denotes the kth power of B_s.

13. Show that for each fixed $t \geq 0$, the approximating sums:

$$\sum_{k=0}^{2^n - 1} B\left(\left(k + \frac{1}{2}\right) t 2^{-n}\right) \{B((k+1)t2^{-n}) - B(kt2^{-n})\}$$

converge in L^2 to $\int_0^t B_s \, dB_s + t/2$ as $n \to \infty$. Note that this limit does *not* define a martingale. It defines what is usually called the *Stratonovitch* integral of B with respect to B.

3

EXTENSION OF THE
PREDICTABLE INTEGRANDS

3.1 Introduction

In this chapter, we show that the definition of the stochastic integral
can be extended to a larger class of integrands than the predictable ones,
when either a mild condition on the Doléans measure μ_M is satisfied or M
is continuous.

First we discuss the relationship of the predictable and optional σ-fields
to the σ-fields generated by various classes of adapted processes.

3.2 Relationship between $\mathcal{P}$, $\mathcal{O}$, and Adapted Processes

Notation. We use the following abbreviations: c. for "continuous", r.c. for
"right continuous", l.c. for "left continuous", r.c.l.l. for "right continuous
with finite left limits". In some papers written in English, the French
equivalent of r.c.l.l. is used, namely, cadlag—"continu à droite et limité à
gauche". The abbreviation l.c.r.l. is defined in the obvious way. We use
$\sigma(\text{c.})$, $\sigma(\text{r.c.})$, etc., to denote the σ-fields of subsets of $I\!\!R_+ \times \Omega$ generated
by the c., r.c., etc., *adapted* processes. For instance, $\sigma(\text{c.})$ is the smallest
σ-field of subsets of $I\!\!R_+ \times \Omega$ with respect to which every continuous adapted

process is $\sigma(\text{c.})$-measurable.

We shall prove that each l.c. adapted process is predictable and the predictable σ-field $\mathcal{P}$ is generated by the continuous adapted processes. On the other hand, we shall sketch the proof that each r.c. adapted process is optional and prove that the optional σ-field $\mathcal{O}$ is generated by the r.c.l.l. adapted processes.

Theorem 3.1. *Any left continuous adapted process is predictable.*

Proof. Suppose X is a left continuous adapted process. For each n let

$$X_s^n = \begin{cases} X_0 & \text{if } s = 0 \\ X_{k2^{-n}} & \text{if } s \in (k2^{-n}, (k+1)2^{-n}] \text{ for } k \in \mathbb{N}_0. \end{cases}$$

Then for any Borel set A in $\mathbb{R}$,

$$(X^n)^{-1}(A) = \left(\{0\} \times X_0^{-1}(A) \right)$$
$$\cup \left(\bigcup_{k=0}^{\infty} \left(k2^{-n}, (k+1)2^{-n} \right] \times X_{k2^{-n}}^{-1}(A) \right).$$

Since X is adapted, the above set is in $\mathcal{P}$, and therefore X^n is predictable. By the left continuity of X, X^n converges pointwise on $\mathbb{R}_+ \times \Omega$ to X, and hence X is predictable. $\blacksquare$

The preceding theorem shows that $\sigma(\text{l.c.}) \subset \mathcal{P}$. Conversely, since the indicator function of a predictable rectangle is a l.c.r.l. adapted process, it follows that $\mathcal{P} \subset \sigma(\text{l.c.r.l.})$. Moreover we have the following.

Theorem 3.2. $\sigma(\text{l.c.r.l.}) \subset \sigma(\text{c.})$.

Proof. It suffices to prove that any bounded l.c.r.l. adapted process X is a pointwise limit on $\mathbb{R}_+ \times \Omega$ of continuous adapted processes. Extend the definition of X_s to negative values of s by setting $X_s = X_0$ for $s < 0$. For each fixed $n \in \mathbb{N}$ and $t \geq 0$, for each $m \in \mathbb{N}$ let

$$(3.1) \qquad \begin{aligned} Y^m(s, \omega) = & 1_{\{t-\frac{1}{n}\}}(s) X_{t-\frac{1}{n}}(\omega) \\ & + \sum_k 1_{(k2^{-m}, (k+1)2^{-m}]}(s) X_{k2^{-m}}(\omega), \end{aligned}$$

where the sum is over all integers k such that $k2^{-m} \in [t - \frac{1}{n}, t]$. Then Y^m is a $\mathcal{B}\left([t - \frac{1}{n}, t]\right) \times \mathcal{F}_t$-measurable function of $(s, \omega) \in [t - \frac{1}{n}, t] \times \Omega$. By the left continuity of X, Y^m converges pointwise on $[t - \frac{1}{n}, t] \times \Omega$ to X as $m \to \infty$, and therefore X inherits the measurability property of the Y^m. Since $X(\cdot, \omega)$ is l.c.r.l.,

$$X_t^n(\omega) \equiv n \int_{t-\frac{1}{n}}^{t} X(s, \omega) \, ds$$

is well defined for each ω. Moreover, by the $\mathcal{B}\left([t - \frac{1}{n}, t]\right) \times \mathcal{F}_t$-measurability of X on $[t - \frac{1}{n}, t] \times \Omega$ and Fubini's theorem, X_t^n is $\mathcal{F}_t$-measurable. Also, the integral defining X_t^n is continuous in t. Hence X^n is a continuous adapted process and by the left continuity of X, X^n converges to X pointwise on $\mathbb{R}_+ \times \Omega$, as desired. ∎

By combining the above results we obtain

(3.2) $$\mathcal{P} = \sigma(\text{c.}) = \sigma(\text{l.c.r.l.}) = \sigma(\text{l.c.}).$$

In Chapter 2 we proved that $\mathcal{P} \subset \mathcal{O}$. By Lemma 2.2, $\mathcal{O}$ is generated by the stochastic intervals of the form $[\tau, \infty)$. The indicator functions of these intervals are r.c.l.l. adapted processes (since τ is optional), consequently $\mathcal{O} \subset \sigma(\text{r.c.l.l.})$. We prove the reverse inclusion below. Furthermore, a modification of the proof shows that $\sigma(\text{r.c.}) \subset \mathcal{O}$. The following is a preliminary lemma.

Lemma 3.3. *Let X be a right continuous adapted process and τ and η be optional times. Then*

(i) $X_\tau 1_{\{\tau < \infty\}} \in \mathcal{F}_\tau$,

(ii)   *For any random variable $Y \in \mathcal{F}_\tau$, the process $Y 1_{[\tau, \eta)}$ is optional.*

Remark. In (i), the value of $Y(\omega)$ is defined to be zero on $\{\omega : \tau(\omega) = \infty\}$, even when X_∞ is not defined.

Proof. Part (i) is proved in Chung [12; Theorem 10, p. 19 and p. 15]. Since any $Y \in \mathcal{F}_\tau$ can be expressed as a pointwise limit of $\mathcal{F}_\tau$-simple functions, it suffices to prove (ii) for $Y = 1_\Lambda$ where $\Lambda \in \mathcal{F}_\tau$. Assuming Y is of this form, define τ_Λ to equal τ on Λ and ∞ on $\Omega \backslash \Lambda$, and define η_Λ similarly. Then $Y 1_{[\tau,\eta)} = 1_{[\tau_\Lambda, \eta_\Lambda)}$. Now τ_Λ and η_Λ are optional times, hence $[\tau_\Lambda, \eta_\Lambda)$ is a stochastic interval and therefore an optional set. Then (ii) follows. ∎

In the proof of the next theorem, we approximate a r.c.l.l. adapted process X uniformly on $I\!R_+ \times \Omega$ by optional processes of the form $\sum_{k=0}^{\infty} X_{\tau_k} 1_{[\tau_k, \tau_{k+1})}$ where $\{\tau_k\}$ is an increasing sequence of optional times chosen so that "X has not moved much during the time between τ_k and τ_{k+1}". It is instructive to compare this with the simpler proof of Theorem 3.1.

Theorem 3.4. $\sigma(\text{r.c.l.l.}) \subset \mathcal{O}$.

Proof. We show that any r.c.l.l. adapted process X is a pointwise limit on $I\!R_+ \times \Omega$ of a sequence of optional processes $\{X^n\}$. For each n, the definition of X^n involves optional times $\tau_k^n, k \in I\!N_0$, defined inductively as follows:

$$\tau_0^n \equiv 0$$

$$\tau_{k+1}^n = \inf \left\{ s > \tau_k^n : \left| X_s - X_{\tau_k^n} \right| > \frac{1}{n} \right\} \quad \text{for } k \in I\!N_0.$$

The inf of an empty set is defined by convention to be ∞. To verify that τ_k^n is optional, we proceed by induction on k. The case $k = 0$ is trivial, so we assume τ_k^n is optional for some $k \geq 0$. Then by the right continuity of X, we have for each t:

$$\{\tau_{k+1}^n < t\} = (\{\tau_k^n < t\})$$

$$\cap \bigcup_s \left(\{\tau_k^n < s\} \cap \left\{ \left| X_s - X_{\tau_k^n \wedge t} \right| > \frac{1}{n} \right\} \right)$$

where the union is over all rationals s in $(0, t)$. By the induction assumption, the adaptedness of X, and since $X_{\tau_k^n \wedge t} \in \mathcal{F}_t$ by Lemma 3.3(i), the above set is in $\mathcal{F}_t$.

Since X has right continuous paths, on the subset of Ω where τ_{k+1}^n is finite we have

$$(3.3) \qquad \left| X_{\tau_{k+1}^n} - X_{\tau_k^n} \right| \geq \frac{1}{n},$$

which implies $\tau_k^n < \tau_{k+1}^n$. It also follows from (3.3) that $\{\tau_k^n, k \in I\!N_0\}$ cannot have a finite limit point; otherwise, immediately to the left of this finite point, the sample path would oscillate by more than $\frac{1}{n}$, contradicting the fact that X has finite left limits. Hence $\tau_k^n \uparrow \infty$ as $k \uparrow \infty$.

For each n, let $X^n = \sum_{k=0}^{\infty} X_{\tau_k^n} 1_{[\tau_k^n, \tau_{k+1}^n)}$. The kth term in this sum equals $X_{\tau_k^n} 1_{\{\tau_k^n < \infty\}} 1_{[\tau_k^n, \tau_{k+1}^n)}$ and by Lemma 3.3(ii) this is an optional process; hence so is the sum X^n. By the definition of the τ_k^n, $|X - X^n| \leq \frac{1}{n}$ for all n and consequently $X = \lim_{n \to \infty} X^n$ pointwise on $I\!R_+ \times \Omega$, as desired. ∎

Remark. The above proof can be modified to show that $\sigma(\text{r.c.}) \subset \mathcal{O}$. Since this result is not needed, its proof will be sketched only. The main difference for right continuous processes is that left limits need not exist, and hence the sequence of optional times $\{\tau_k^n, k \in I\!N_0\}$ defined above can have a finite limit point. Consequently, transfinite induction must be used to define for each n a family of optional times $\{\tau_\alpha^n\}$, indexed by the countable ordinals α, such that $\tau_0^n \equiv 0$,

$$\tau_{\alpha+1}^n = \inf\left\{s > \tau_\alpha^n : \left|X_s - X_{\tau_\alpha^n}\right| > \frac{1}{n}\right\}$$

for every countable ordinal α, and $\tau_\alpha^n = \sup_{\beta < \alpha} \tau_\beta^n$ when α is a limit ordinal. Those τ_α^n which are finite-valued are strictly increasing with α and there is a countable ordinal α_1, independent of ω, such that $\tau_{\alpha_1}^n = \infty$. With these changes to the proof of Theorem 3.4, it follows that $\sigma(\text{r.c.}) \subset \mathcal{O}$.

If X is a right continuous adapted process, then $X = \lim_{n \to \infty} X^n$ pointwise on $I\!R_+ \times \Omega$ where

$$X^n(s, \omega) = \sum_{k=0}^{\infty} X_{(k+1)2^{-n}}(\omega) 1_{[k2^{-n}, (k+1)2^{-n})}(s),$$

defines a $\mathcal{B} \times \mathcal{F}$-measurable process for each n. It follows that $\sigma(\text{r.c.}) \subset \mathcal{B} \times \mathcal{F}$.

Another class of adapted processes will be discussed in the next section, in connection with the extension of stochastic integrals with respect to continuous L^2-martingales. This is the class of progressively measurable processes. A function $Z : I\!R_+ \times \Omega \to I\!R$ is a *progressively measurable process* if and only if for each t the restriction of Z to $[0,t] \times \Omega$ is $\mathcal{B}_t \times \mathcal{F}_t$-measurable. The σ-field of subsets of $\mathbb{R}_+ \times \Omega$ generated by the progressively measurable (or progressive) processes will be denoted by $\mathcal{M}$. Each progressive process is clearly $\mathcal{B} \times \mathcal{F}$-measurable and by Fubini's theorem, each is adapted. All optional processes are progressive. To see this, it suffices to show that any right continuous adapted process X is progressive. But for each fixed t, on $[0,t] \times \Omega$, such an X is the pointwise limit of the $\mathcal{B}_t \times \mathcal{F}_t$-measurable processes X^n defined by

$$X^n(s,\omega) = \sum_{k=0}^{2^n-1} X\left(\frac{(k+1)t}{2^n},\omega\right) 1_{[kt2^{-n},(k+1)t2^{-n})}(s) + X(t,\omega)1_{\{t\}}(s),$$

where $(s,\omega) \in [0,t] \times \Omega$. On the other hand, there are $\mathcal{B} \times \mathcal{F}$-measurable adapted processes which are not progressively measurable. The following probabilistic example, communicated to us by Martin Barlow, illustrates this.

Example. Let $B = \{B(t),\ t \in \mathbb{R}_+\}$ be a one-dimensional Brownian motion For definiteness we suppose B starts from zero. Let $\{\mathcal{F}_t,\ t \in \mathbb{R}_+\}$ be the associated standard filtration (cf. Section 1.8). Define

$$S = \sup\{t \in [0,1] : B(t) = 0\},$$

i.e., S is the *last* time B is at zero before time one. Then S is an $\mathcal{F}$-measurable random variable, but it is *not* an optional time (why?). It follows that the graph of S, $\{(S(\omega), \omega) : \omega \in \Omega\}$, is $\mathcal{B} \times \mathcal{F}$-measurable and hence so too is the indicator function X of this set. Moreover, X is adapted because

$$P(X_t = 1) = P(S = t) = 0 \quad \text{for all}\quad t \geq 0,$$

and $\mathcal{F}_t$ contains all of the P-null sets in $\mathcal{F}$. On the other hand, if X were progressive, then for each $t \in \mathbb{R}_+$,

$$X^{-1}(1) \bigcap ([0,t] \times \Omega) = \{(S(\omega),\omega) : \omega \in \Omega,\ S(\omega) \in [0,t]\}$$

would be $\mathcal{B}_t \times \mathcal{F}_t$-measurable, and then by the theory of analytic sets (see Dellacherie and Meyer [23; III-13, III-33]), its projection $\{\omega : S(\omega) \leq t\}$ on Ω would be an analytic set and hence $\mathcal{F}_t$-measurable because $\mathcal{F}_t$ is complete.

But for $t \in (0,1)$ this is false, from the definition of S. Thus, X is a $\mathcal{B} \times \mathcal{F}$-measurable adapted process, but it is not progressively measurable.

For other properties of progressively measurable processes, see Chung [12, Section 1.5] or Dellacherie and Meyer [23; IV-14, IV-50ff.].

In summary, we have the following relationships between the σ-fields discussed above:

$$\mathcal{P} = \sigma(\mathrm{c.}) = \sigma(\mathrm{l.c.r.l.}) = \sigma(\mathrm{l.c.})$$
$$(3.4) \qquad \subset \mathcal{O} = \sigma(\mathrm{r.c.l.l.}) = \sigma(\mathrm{r.c.})$$
$$\subset \mathcal{M} \subset \mathcal{B} \times \mathcal{F}.$$

3.3 Extension of the Integrands

In this section, we discuss two conditions, under either of which the definition of the stochastic integral can be extended to an augmented class of predictable integrands. For this we shall assume that M is a right continuous L^2-martingale. The extension (by localization) to right continuous local L^2-martingale integrators is left to the reader (cf. Section 2.6). The first condition requires the Doléans measure μ_M to be absolutely continuous with respect to the product measure $\lambda \times P$. In this case the augmented class includes suitably integrable $\mathcal{B} \times \mathcal{F}$-measurable adapted processes. The alternative condition is that M be continuous. In this case the augmented class includes suitably integrable progressive processes. If M is a Brownian motion, both conditions are satisfied and the extensions coincide.

For consideration of the first condition, until stated otherwise, we make the following assumption.

Assumption A.1. Suppose M is a right continuous L^2-martingale such that its Doléans measure μ_M is absolutely continuous with respect to $\lambda \times P$ on $\mathcal{P}$, i.e., $\mu_M \ll \lambda \times P$.

Remark. The extension under this assumption can also be applied to a right continuous local L^2-martingale M such that $M - M_0$ satisfies the assumption, because stochastic integrals with respect to M are the same as those with respect to $M - M_0$ (see Section 2.6).

Example. A Brownian motion B in $I\!R$ that starts from zero satisfies the above assumption with $\mu_B = \lambda \times P$. For a general Brownian motion in $I\!R$, $B - B_0$ satisfies the above assumption.

Since $\mu_M \ll \lambda \times P$ on $\mathcal{P}$, by the Radon-Nikodym theorem (see Royden [69, p. 238]), there is a $\mathcal{P}$-measurable function f such that $0 \le f < \infty$ and for each $A \in \mathcal{P}$:

$$(3.5) \qquad \mu_M(A) = \int_A f \, d(\lambda \times P).$$

The right member of (3.5) actually defines a measure on sets A in $\mathcal{B} \times \mathcal{F}$. This extension of μ_M will be denoted by $\tilde{\mu}_M$.

Definition. Let $\mathcal{N}^*$ denote the class of all $\lambda \times P$-null sets in $\mathcal{B} \times \mathcal{F}$. The augmentation of $\mathcal{P}$ with respect to $(\mathcal{B} \times \mathcal{F}, \lambda \times P)$ is the σ-field of subsets of $I\!R_+ \times \Omega$ generated by $\mathcal{P}$ and $\mathcal{N}^*$. We denote it by $\mathcal{P}^*$. Similarly, we define $\tilde{\mathcal{N}}$ to be the class of $\tilde{\mu}_M$-null sets in $\mathcal{B} \times \mathcal{F}$ and $\tilde{\mathcal{P}}$ to be the σ-field generated by $\mathcal{P}$ and $\tilde{\mathcal{N}}$. Note that $\mathcal{N}^* \subset \tilde{\mathcal{N}}$ and hence $\mathcal{P}^* \subset \tilde{\mathcal{P}}$.

Remark. For the extension of the stochastic integral under Assumption A.1, we only need the larger σ-field $\tilde{\mathcal{P}}$. However, to elucidate the measurability properties of certain integrands (cf. Theorem 3.7), we also consider the σ-field $\mathcal{P}^*$. Note that the definition of this σ-field does not depend on M.

The sets in $\mathcal{P}^*$ (or $\tilde{\mathcal{P}}$) and the associated measurable functions are characterized by the following standard result, which can be obtained by replacing Ω, G, and G^*, in Chung [12, pp. 59–60] by $I\!R_+ \times \Omega$, $\mathcal{P}$, and $\mathcal{P}^*$ (or $\tilde{\mathcal{P}}$), respectively.

Lemma 3.5.

(i) $\quad \mathcal{P}^* = \{A \in \mathcal{B} \times \mathcal{F} : A \triangle A_1 \in \mathcal{N}^* \text{ for some } A_1 \in \mathcal{P}\}.$

(ii) *Suppose $Z : I\!R_+ \times \Omega \to I\!R$ is $\mathcal{B} \times \mathcal{F}$-measurable. Then Z is $\mathcal{P}^*$-measurable if and only if there exists a predictable process X such that $(\lambda \times P)(X \neq Z) = 0$.*

(iii) *The results (i) and (ii) also hold with $\tilde{\mathcal{N}}$, $\tilde{\mathcal{P}}$ and $\tilde{\mu}_M$ in place of $\mathcal{N}^*$, $\mathcal{P}^*$ and $\lambda \times P$, respectively.*

We shall now prove that all measurable adapted processes are $\mathcal{P}^*$-measurable. Before doing this we establish the *optional projection theorem* (cf. Dellacherie and Meyer [24, pp. 113–114]). A simple form of this is needed for the proof of Theorem 3.7 and the full force of it will be needed later to prove Theorem 3.10.

Theorem 3.6. *(Optional Projection Theorem.) For any bounded $\mathcal{B} \times \mathcal{F}$-measurable process Z, there is an optional process Y such that for each optional time τ,*

$$(3.6) \qquad E[Z_\tau 1_{\{\tau < \infty\}} \mid \mathcal{F}_\tau] = Y_\tau 1_{\{\tau < \infty\}} \qquad \text{P-a.s.}$$

Remark. The process Y is in fact unique, up to indistinguishability (see [24, p. 114]).

Proof. For a proof via a monotone class argument, let $\mathcal{Z}$ denote the class of all bounded $\mathcal{B} \times \mathcal{F}$-measurable Z for which there is such a Y. The collection of sets of the form $[0, s) \times F$, where $s \in I\!\!R_+$ and $F \in \mathcal{F}$, is closed under finite intersections. First we show that the indicator function of any set in this collection is in $\mathcal{Z}$. Let $s \in I\!\!R_+$, $F \in \mathcal{F}$, and $Z = 1_{[0, s) \times F}$. By [12; p. 30, 26, 32], there is a r.c.l.l. version N of the uniformly integrable martingale $\{E(1_F \mid \mathcal{F}_t), t \in I\!\!R_+\}$ and for each optional time τ, $E[1_F \mid \mathcal{F}_\tau] = N_\tau$ P-a.s. on $\{\tau < \infty\}$. It follows that $Y = 1_{[0, s) \times \Omega} N$ satisfies (3.6), and since it is a r.c.l.l. adapted process, it is optional, by Theorem 3.4. Thus $1_{[0,s) \times F} \in \mathcal{Z}$. Clearly $\mathcal{Z}$ is closed under finite linear combinations. Moreover, if $\{Z^n\}$ is a bounded monotone sequence in $\mathcal{Z}$ with corresponding optional processes $\{Y^n\}$ and limit Z, then $Y \equiv \liminf_n Y^n 1_{\{|\liminf_n Y^n| < \infty\}}$ is an optional process satisfying (3.6), by the monotone convergence of conditional expectations. Thus $\mathcal{Z}$ is closed under bounded monotone limits. It then follows by a monotone class theorem for functions (see Section 1.2) that $\mathcal{Z}$ contains all bounded $\mathcal{B} \times \mathcal{F}$-measurable processes, proving the desired result. ∎

Theorem 3.7. *Any $\mathcal{B} \times \mathcal{F}$-measurable adapted process is $\mathcal{P}^*$-measurable.*

Proof. First we prove $\mathcal{O} \subset \mathcal{P}^*$, from which it follows that any optional process is $\mathcal{P}^*$-measurable. By Lemma 2.2, $\mathcal{O}$ is generated by the stochastic intervals of the form $[\tau, \infty)$, and by Lemma 2.1, the stochastic interval (τ, ∞) is in $\mathcal{P} \subset \mathcal{P}^*$. Thus to prove $\mathcal{O} \subset \mathcal{P}^*$, it suffices to show that $[\tau] \in \mathcal{P}^*$ for each optional time τ. Now $[\tau]$, being a stochastic interval, is contained in $\mathcal{O}$, and so by (3.4), $[\tau] \in \mathcal{B} \times \mathcal{F}$. Also, $\lambda\{t : \tau(\omega) = t\} = 0$ for each ω. Thus by Fubini's theorem, $(\lambda \times P)([\tau]) = 0$. Hence $[\tau] \in \mathcal{N}^* \subset \mathcal{P}^*$ and it follows that $\mathcal{O} \subset \mathcal{P}^*$.

Now, to prove Theorem 3.7, it suffices to consider a bounded $\mathcal{B} \times \mathcal{F}$-measurable adapted process Z. Let Y be an optional process satisfying (3.6), so that in particular, $Y_t = E[Z_t \mid \mathcal{F}_t]$ for each $t \in \mathbb{R}_+$. Since Z is adapted, Z_t is a member of the equivalence class $E(Z_t \mid \mathcal{F}_t)$, and so $Y_t = Z_t$ P-a.s. for each $t \in \mathbb{R}_+$. It then follows from Fubini's theorem (note Y and Z are $\mathcal{B} \times \mathcal{F}$-measurable), that $(\lambda \times P)(Y \neq Z) = 0$. Since, Y is optional, by the first part of this proof, $Y \in \mathcal{P}^*$, and then by Lemma 3.5(ii), there is $X \in \mathcal{P}$ such that $(\lambda \times P)(X \neq Y) = 0$. Hence, $(\lambda \times P)(X \neq Z) = 0$ and by applying Lemma 3.5(ii) again, we conclude $Z \in \mathcal{P}^*$. ∎

To summarize the measurability properties under Assumption A.1, let $\mathcal{V}$ denote the σ-field generated by the $\mathcal{B} \times \mathcal{F}$-measurable adapted processes. Then we have

$$\mathcal{P} \subset \mathcal{O} \subset \mathcal{M} \subset \mathcal{V} \subset \mathcal{P}^* \subset \tilde{\mathcal{P}} \subset \mathcal{B} \times \mathcal{F}.$$

We now show how to extend stochastic integrals with respect to M to suitable $\tilde{\mathcal{P}}$-measurable integrands.

Notation. For any sub-σ-field $\mathcal{W}$ of $\mathcal{B} \times \mathcal{F}$, we shall use $\mathcal{L}^2_{\mathcal{W}}$ to denote $L^2(\mathbb{R}_+ \times \Omega, \mathcal{W}, \tilde{\mu}_M)$ and $\Lambda(\mathcal{W}, M)$ to denote the set of $\mathcal{W}$-measurable processes X for which there is a sequence of optional times $\{\tau_k\}$ such that $\tau_k \uparrow \infty$ a.s. and $1_{[0, t \wedge \tau_k]} X \in \mathcal{L}^2_{\mathcal{W}}$ for each t and k. Such a sequence $\{\tau_k\}$ will be called a localizing sequence for (X, M).

Since $\mathcal{P} \subset \tilde{\mathcal{P}}$, then $\mathcal{L}^2_{\mathcal{P}} \subset \mathcal{L}^2_{\tilde{\mathcal{P}}}$. Conversely, by Lemma 3.5(iii), for any

$Z \in \mathcal{L}_{\tilde{\mathcal{P}}}^2$ there is $X \in \mathcal{L}_{\mathcal{P}}^2$ such that $X = Z$, $\tilde{\mu}_M$-a.e. It follows that as Hilbert spaces $\mathcal{L}_{\mathcal{P}}^2$ and $\mathcal{L}_{\tilde{\mathcal{P}}}^2$ are the same. For $X \in \mathcal{L}_{\mathcal{P}}^2$, $\int X\, dM$ was defined by the isometry between $\mathcal{L}_{\mathcal{P}}^2$ and L^2. Consequently, $\int Z\, dM$ is defined by this isometry for any $Z \in \mathcal{L}_{\tilde{\mathcal{P}}}^2$. Then $\int_{[0,t]} Z\, dM$ can be defined by the usual extension procedure for any $Z \in \Lambda(\tilde{\mathcal{P}}, M)$. Indeed, for $Z \in \Lambda(\tilde{\mathcal{P}}, M)$ and $X \in \mathcal{P}$ such that $\tilde{\mu}_M(X \neq Z) = 0$, we have almost surely for all t:

$$\int_{[0,t]} Z\, dM = \int_{[0,t]} X\, dM.$$

The following theorem gives a convenient sufficient condition for a process to be in $\Lambda(\tilde{\mathcal{P}}, M)$ when M is a Brownian motion. Note from the remark at the beginning of this section that it suffices to consider a Brownian motion starting from zero. In this case, since $\mu_B = \lambda \times P$, we have $\tilde{\mathcal{P}} = \mathcal{P}^*$.

Theorem 3.8. *Suppose Z is a $\mathcal{B} \times \mathcal{F}$-measurable adapted process such that we have a.s. for all t:*

$$(3.7) \qquad \int_0^t (Z_s)^2\, ds < \infty.$$

If B is a Brownian motion in $\mathbb{R}$ that starts from zero, then $Z \in \Lambda(\tilde{\mathcal{P}}, B)$.

Proof. By Theorem 3.7 and Lemma 3.5(ii), there is $X \in \mathcal{P}$ such that $X = Z$, $(\lambda \times P)$-a.e. Then by Fubini's theorem, for almost every ω, $X(\,\cdot\,, \omega) = Z(\,\cdot\,, \omega)$, λ-a.e. Consequently, a.s., (3.7) holds for all t with X in place of Z. In fact, since $\mathcal{F}_0$ contains all of the P-null sets, we can modify X on a set of the form $[0, \infty) \times F$, where F is a P-null set, to ensure that

$$\int_0^t (X(s, \omega))^2\, ds < \infty \quad \text{for all } (t, \omega) \in \mathbb{R}_+ \times \Omega,$$

whilst preserving the properties that $X \in \mathcal{P}$ and $X = Z$, $(\lambda \times P)$-a.e. Then, by considering the case where X is $\mathcal{R}$-simple and using a monotone class argument, it follows that $\int_0^t (X_s)^2\, ds$ is $\mathcal{F}_t$-measurable for each t. Moreover this integral is continuous in t. Define

$$\tau_k = \inf \left\{ t \geq 0 : \int_0^t (X_s)^2\, ds \geq k \right\}, \quad \text{for } k = 1, 2, \ldots.$$

Then $\{\tau_k\}$ is a non-decreasing sequence of optional times satisfying $\tau_k \to \infty$ as $k \to \infty$. Moreover, for each t and k, we have a.s.

$$\int_0^{t \wedge \tau_k} (Z_s)^2 \, ds = \int_0^{t \wedge \tau_k} (X_s)^2 \, ds \le k.$$

Taking expectations, it follows that

$$1_{[0, t \wedge \tau_k]} Z \in L^2(\mathbb{R}_+ \times \Omega, \tilde{\mathcal{P}}, \lambda \times P)$$

for each t and k. Hence $Z \in \Lambda(\tilde{\mathcal{P}}, B)$. ∎

For the remainder of this section, we shall assume that the following holds in place of Assumption A.1.

Assumption A.2. Suppose that M is a continuous L^2-martingale.

For the discussion of this case, we need the quadratic variation process associated with M, which is defined in the next chapter. However, for our purposes here, we only need the following proposition which is immediate from Theorem 4.2.

Proposition 3.9. Let M be a continuous L^2-martingale. Then the quadratic variation process $[M]$ of M is a continuous, adapted, increasing process satisfying

$$(3.8) \qquad \mu_M(A) = E \left(\int_0^\infty 1_A(s, \omega) \, d[M]_s(\omega) \right),$$

for each set $A \in \mathcal{P}$.

Remark. Here, for each $\omega \in \Omega$,

$$\int_0^\infty 1_A(s, \omega) \, d[M]_s(\omega)$$

is defined as a Lebesgue–Stieltjes integral.

Example. If M is a Brownian motion, then $[M]_t = t$ for all $t \in \mathbb{R}_+$.

The right member of (3.8) defines a measure on all sets $A \in \mathcal{B} \times \mathcal{F}$. We shall denote this extension of μ_M by $\tilde{\mu}_M$. The augmentation of $\mathcal{P}$

and the associated extension of the stochastic integral is then defined in an analogous manner to that under Assumption A.1. Specifically, let $\tilde{N}$ denote the collection of $\tilde{\mu}_M$-null sets in $\mathcal{B} \times \mathcal{F}$, and define $\tilde{\mathcal{P}} = \mathcal{P} \vee \tilde{N}$. Then, as in the discussion preceding Theorem 3.8, $\int_{[0,t]} Z\, dM$ can be defined for all $Z \in \Lambda(\tilde{\mathcal{P}}, M)$.

The following theorem indicates that progressive processes are in the augmentation $\tilde{\mathcal{P}}$ of $\mathcal{P}$. The proof given here was indicated to us by Michael Sharpe.

Theorem 3.10. *Any progressively measurable process is $\tilde{\mathcal{P}}$-measurable.*

Proof. First we prove $\mathcal{O} \subset \tilde{\mathcal{P}}$. As in the proof of Theorem 3.7, it suffices to prove the stochastic interval $[\tau] \in \tilde{\mathcal{P}}$ for any optional time τ. But, for any such τ,

$$\tilde{\mu}_M([\tau]) = E\left(\int_0^\infty 1_{[\tau]}(s, \omega)\, d[M]_s(\omega)\right)$$
$$= E\left([M]_\tau - [M]_{\tau-}\right) = 0,$$

since M is continuous.

To prove the theorem, it suffices to consider a bounded progressively measurable process Z. By the optional projection theorem, there is an optional process Y satisfying (3.6) for all optional times τ. Since Z is progressive and $\{\mathcal{F}_t\}$ is right continuous, $Z_\tau 1_{\{\tau < \infty\}} \in \mathcal{F}_\tau$ for all such τ (see Chung [12, p. 39]). Hence

(3.9) $$Z_\tau 1_{\{\tau < \infty\}} = Y_\tau 1_{\{\tau < \infty\}} \quad P\text{-a.s.}$$

Now,

$$\tilde{\mu}_M(Y \neq Z) = E\left(\int_0^\infty 1_{\{Y(s) \neq Z(s)\}} d[M]_s\right).$$

To evaluate the right member above, we perform a change of variable in the integral over $\mathbb{R}_+$. For each $t \geq 0$, define

$$\tau_t = \inf\{s \geq 0 : [M]_s > t\}.$$

Then $\{\tau_t, t \geq 0\}$ is the right continuous inverse of $[M]$ and for each t, τ_t is an optional time. After performing the change of variable we obtain (cf. Dellacherie and Meyer [24, p. 132]),

$$\int_0^\infty 1_{\{Y(s) \neq Z(s)\}} \, d[M]_s = \int_{[0,[M]_\infty)} 1_{\{Y(\tau_s) \neq Z(\tau_s)\}} \, ds$$
$$= \int_0^\infty 1_{\{\tau_s < \infty, \, Y(\tau_s) \neq Z(\tau_s)\}} \, ds.$$

Hence, using Fubini's theorem and (3.9), we have

$$\tilde{\mu}_M(Y \neq Z) = E \left(\int_0^\infty 1_{\{\tau_s < \infty, \, Y(\tau_s) \neq Z(\tau_s)\}} \, ds \right)$$
$$= \int_0^\infty P(\tau_s < \infty, \, Y(\tau_s) \neq Z(\tau_s)) \, ds$$
$$= 0.$$

Since Y is optional, $Y \in \tilde{\mathcal{P}}$, by the first part of this proof. Then, by Lemma 3.5(iii), there is $X \in \mathcal{P}$ such that $\tilde{\mu}_M(X \neq Y) = 0$ and hence $\tilde{\mu}_M(X \neq Z) = 0$. By applying Lemma 3.5(iii) again, we conclude $Z \in \tilde{\mathcal{P}}$. ∎

Remark.   An alternative proof of the above result, which does not go through optional processes, can be obtained by using the predictable projection theorem (cf. Dellacherie and Meyer [24, pp. 113–114]) and the predictable time change $\tau_t = \inf\{s \geq 0 : [M]_s \geq t\}$.

Summarizing the measurability under Assumption A.2, we have

$$\mathcal{P} \subset \mathcal{O} \subset \mathcal{M} \subset \tilde{\mathcal{P}} \subset \mathcal{B} \times \mathcal{F}.$$

Now, $\mathcal{M} \subset \mathcal{V}$ (recall $\mathcal{V}$ is generated by the $\mathcal{B} \times \mathcal{F}$-measurable adapted processes), and when M is a Brownian motion, $\mathcal{V} \subset \tilde{\mathcal{P}}$. However, the latter relation does not hold in general for a continuous L^2-martingale M. We illustrate this with the following example, due to Martin Barlow, of a continuous L^2-martingale M and a $\mathcal{B} \times \mathcal{F}$-measurable adapted process X such that $X \notin \tilde{\mathcal{P}}$. Although this example uses the local time of a one-dimensional Brownian motion at zero, which is defined in Chapter 7, sufficient details are given here that the idea of the example can be appreciated without knowing the exact definition or all the properties of local time.

Example. Let B and $\tilde{B}$ be two independent one-dimensional Brownian motions starting from zero. Let L be the local time of B at zero. In particular, L is a continuous, increasing process, adapted to B, and L increases only when B is at zero, i.e., $\int_0^t 1_{\{B(s)=0\}} \, dL(s) = L(t)$ for all $t \geq 0$. Let $\mathcal{F}$ be the completion of the σ-field generated by B and $\tilde{B}$, and for each $t \geq 0$, let $\mathcal{F}_t$ denote the augmentation by the P-null sets in $\mathcal{F}$ of the σ-field generated by B and $\{\tilde{B}(s) : 0 \leq s \leq t\}$. For each $t \geq 0$, $L(t)$ is an optional time relative to the filtration $\{\mathcal{F}_s, s \geq 0\}$. Define $M_t = \tilde{B}_{L(t)}$ for each $t \geq 0$. Then it can be shown that $M = \{M_t, \mathcal{F}_{L(t)}, t \geq 0\}$ is a continuous L^2-martingale with quadratic variation process L. Let Y be a bounded $\mathcal{F}$-measurable random variable and define $X(t) = Y 1_{\{s>0:B(s)=0\}}(t)$, for each $t \geq 0$. Then X is a $\mathcal{B} \times \mathcal{F}$-measurable process which is adapted to $\{\mathcal{F}_t\}$ since $X(t) = 0$ P-a.s. for each $t \geq 0$ and $\mathcal{F}_t$ contains all of the P-null sets. Now (cf. (3.8)), $\tilde{\mu}_M(X \neq Y) = 0$, because

$$\tilde{\mu}_M(\{(s, \omega) : B(s, \omega) \neq 0\}) = E\left(\int_0^\infty 1_{\{B(s)\neq 0\}} dL(s)\right) = 0$$

and $\tilde{\mu}_M([0]) = 0$. Thus, if X were in $\tilde{\mathcal{P}}$ then Y would be also (cf. Lemma 3.5), and $\int_0^t X \, dM$ would equal $\int_0^t Y \, dM = Y(M_t - M_0)$. But the latter is not adapted in general, let alone a local martingale (take $Y = M_1$ for example). Hence $X \notin \tilde{\mathcal{P}}$.

3.4 A Historical Note

For stochastic integration with respect to Brownian motion, Itô [46, p. 175 ff.] took his integrands to be the $\mathcal{B} \times \mathcal{F}$-measurable, adapted processes satisfying (3.7). He used the term "non-anticipating" in place of "adapted". The essential result which allowed the use of these integrands was Theorem 1 (loc. cit., p. 176). There Itô proved that if Z is an integrand which is bounded on $[0, t] \times \Omega$, then $1_{[0,t]\times\Omega} Z$ is a limit in $\mathcal{L}^2_{\mathcal{B}\times\mathcal{F}}$ of r.c.l.l. adapted processes. Some later authors, e.g., [57, p. 23], however, gave Itô's result without verifying that the approximating processes used were adapted. Part (i) of the following lemma fills this gap. Part (ii), which appears in Letta [52], provides an almost converse of (i).

Lemma   3.11.   *Suppose Z is a $\mathcal{B} \times \mathcal{F}$-measurable process such that $\int_0^t |Z(s,\omega)|\, ds < \infty$ for all $(t,\, \omega) \in \mathbb{R}_+ \times \Omega$.*

(i)   *If Z is adapted, then $\int_0^t Z(s, \cdot)\, ds$ is $\mathcal{F}_t$-measurable for all $t \in \mathbb{R}_+$.*

(ii)   *Conversely, if $\int_0^t Z(s, \cdot)\, ds$ is $\mathcal{F}_t$-measurable for each $t \in \mathbb{R}_+$, then there is a $\mathcal{B} \times \mathcal{F}$-measurable, adapted (in fact, predictable) process X such that $X = Z$, $(\lambda \times P)$-a.e.*

Proof. For the proof of (i), suppose that Z is adapted. Then by Theorem 3.7 and Lemma 3.5, there is $X \in \mathcal{P}$ such that $X = Z$, $(\lambda \times P)$-a.e., and consequently by Fubini's theorem, for each fixed $t \in \mathbb{R}_+$ we have

$$(3.10) \qquad \int_0^t X(s,\omega)\, ds = \int_0^t Z(s,\omega)\, ds \quad P\text{-a.s.}$$

In particular, the left member of (3.10) is well defined P-a.s. Since $X \in \mathcal{P}$, by first considering the case where X is $\mathcal{R}$-simple and then using a monotone class argument, it follows that the left member of (3.10) is P-a.s. equal to an $\mathcal{F}_t$-measurable r.v., and hence so is the right member. Since the right member is well defined everywhere and $\mathcal{F}_t$ contains all of the P-null sets in $\mathcal{F}$, the conclusion of (i) follows.

For the proof of (ii), suppose for each $t \in \mathbb{R}_+$ that the random variable $Y(t, \cdot) \equiv \int_0^t Z(s, \cdot)\, ds$ is $\mathcal{F}_t$-measurable. Extend the definition of $Y(t, \cdot)$ to negative values of t by setting $Y(t, \cdot) \equiv 0$ for all $t < 0$. In the following, h runs through the negative *rational* numbers. For each $(t, \omega) \in \mathbb{R}_+ \times \Omega$, set $X(t, \omega)$ equal to

$$\lim_{h \uparrow 0} \frac{Y(t, \omega) - Y(t - h, \omega)}{h},$$

whenever this limit exists and is finite, and set $X(t, \omega)$ equal to 0 otherwise. Then, since $(t, \omega) \to (Y(t, \omega) - Y(t-h, \omega))/h$ is a continuous adapted process, it is predictable. It follows that the limit process X is also predictable. By Lebesgue's differentiation theorem [69, p. 103], for each ω, $X(\cdot, \omega) = Z(\cdot, \omega)$ λ-a.e. Since X and Z are both $\mathcal{B} \times \mathcal{F}$-measurable, it then follows from Fubini's theorem that $(\lambda \times P)(X \neq Z) = 0$. Since every predictable process is $\mathcal{B} \times \mathcal{F}$-measurable and adapted, this concludes the proof of (ii). ∎

If Z is progressive, the conclusion of Lemma 3.11(i) is an immediate consequence of Fubini's theorem. A more extensive discussion, with and without probability, of the relationship between adapted, progressive, and predictable processes satisfying the hypothesis of Lemma 3.11 is given in the note [52] by Letta. Although a number of general results for $\mathcal{P}$ and $\mathcal{O}$ extend to $\mathcal{M}$, we shall not use these in this monograph.

3.5 Exercises

1. Let $0 \leq s < t < \infty$ and $F \in \mathcal{F}_s$. For $0 \leq s < s' < t < t' < \infty$, let f be the real-valued function on $\mathbb{R}_+$ defined by

$$
f(u) = \begin{cases}
0 & \text{for } 0 \leq u \leq s \\
(u-s)/(s'-s) & \text{for } s < u < s' \\
1 & \text{for } s' \leq u \leq t \\
(t'-u)/(t'-t) & \text{for } t < u < t' \\
0 & \text{for } u \geq t'.
\end{cases}
$$

Define X on $\mathbb{R}_+ \times \Omega$ by $X(u,\omega) = f(u)1_F(\omega)$ for all $(u,\omega) \in \mathbb{R}_+ \times \Omega$. Prove that X is adapted. Similarly, for $F_0 \in \mathcal{F}_0$, $\varepsilon > 0$ and g the real-valued function on $\mathbb{R}_+$ defined by

$$
g(u) = \begin{cases}
(\varepsilon - u)/\varepsilon & \text{for } 0 \leq u \leq \varepsilon \\
0 & \text{for } u > \varepsilon,
\end{cases}
$$

show that Y defined on $\mathbb{R}_+ \times \Omega$ by $Y(u,\omega) = g(u)1_{F_0}(\omega)$ for all $(u,\omega) \in \mathbb{R}_+ \times \Omega$, is adapted. Use these results to give an easy proof that $\mathcal{P} \subset \sigma(c.)$.

2. Let S be an optional time with respect to $\{\mathcal{F}_t\}$ and define

$$
\mathcal{G}_t = \mathcal{F}_{S+t} \quad \text{for all} \quad t \geq 0.
$$

Show that $\{\mathcal{G}_t\}$ is a standard filtration. Let $T : \Omega \to [0, \infty]$ such that $T \geq S$. Prove that T is optional with respect to $\{\mathcal{F}_t\}$ if and only if $1_{\{S<\infty\}}(T - S)$ is optional with respect to $\{\mathcal{G}_t\}$. Use this result to show that each τ_k^n in the proof of Theorem 3.4 is optional.

3. Show that the random variable S defined in the example in Section 3.2 is *not* an optional time and satisfies $P(S = t) = 0$ for all $t \geq 0$.

4

QUADRATIC VARIATION PROCESS

4.1 Introduction

For the remainder of this book, we shall only consider integrators M which are *continuous* local martingales. By Proposition 1.9 these are automatically local L^2-martingales. A more extensive treatment, encompassing right continuous integrators would require more elaborate considerations which are not suitable for inclusion in this short book.

In this chapter, we introduce the quadratic variation process associated with a continuous local martingale. This process plays an important role in the development in Chapter 5 of the Itô formula.

4.2 Definition and Characterization of Quadratic Variation

Definition. For $t \in \mathbb{R}_+$, a *partition* π_t of $[0, t]$ is a finite ordered subset $\pi_t = \{t_0, t_1, \ldots, t_k\}$ of $[0, t]$ such that $0 = t_0 < t_1 < \ldots < t_k = t$. We denote the *mesh* of π_t by

$$\delta\pi_t \equiv \max\{|t_{j+1} - t_j|, j = 0, 1, \ldots, k-1\}.$$

If $\{\pi_t^n, n \in I\!N\}$ is a sequence of partitions of $[0, t]$, then for each n, let t_j^n, $j = 0, 1, \ldots, k_n$ denote the members of π_t^n. To alleviate the printing, we shall omit the superscript n below wherever the meaning is clear from the context.

The main result of this chapter is the following theorem.

Theorem 4.1. *Let* $t \in I\!R_+$ *and* $\{\pi_t^n, n \in I\!N\}$ *be a sequence of partitions of* $[0, t]$ *such that* $\lim_{n \to \infty} \delta\pi_t^n = 0$. *Suppose* M *is a continuous local martingale and for each* n *let*

$$S_t^n = \sum_j \left(M_{t_{j+1}} - M_{t_j} \right)^2$$

where the sum is over all j *such that both* t_j *and* t_{j+1} *are in* π_t^n. *Then*

(i) *if* M *is bounded,* $\{S_t^n, n \in I\!N\}$ *converges in* L^2 *to*

(4.1) $$[M]_t \equiv (M_t)^2 - (M_0)^2 - 2 \int_0^t M \, dM,$$

(ii) $\{S_t^n, n \in I\!N\}$ *converges in probability to* $[M]_t$.

We call $[M]_t$, defined by (4.1), the *quadratic variation* of M at time t, and $[M] = \{[M]_t, t \in I\!R_+\}$ the *quadratic variation process* associated with M.

Remark. In the above, it is essential that the partitions π_t^n do not depend on ω. Indeed, otherwise the supremum of the quadratic sums defining S_t^n over all partitions depending on ω may be $+\infty$.

Before proving Theorem 4.1 we give an important example.

Example. In the case where M is a Brownian motion B in $I\!R$, the quadratic variation process $[B]$ is indistinguishable from $\{t, t \in I\!R_+\}$.

Proof. Since $[B]$ and $\{t, t \in I\!R_+\}$ are continuous processes, to prove they are indistinguishable it suffices to show that $[B]_t = t$ a.s. for each t. For this we exhibit a partition of $[0, t]$ such that $\{S_t^n, n \in I\!N\}$ converges in L^2 to t. For fixed n, let

$$t_j = \frac{jt}{n} \text{ for } j = 0, 1, \ldots, n, \text{ and}$$

$$\Delta_j = (B(t_{j+1}) - B(t_j))^2 - \frac{t}{n} \text{ for } j = 0, 1, \ldots, n-1.$$

Then,

$$E\left\{(S_t^n - t)^2\right\} = E\left\{\left(\sum_{j=0}^{n-1} \Delta_j\right)^2\right\}$$

(4.2)

$$= \sum_{j=0}^{n-1} E\left\{(\Delta_j)^2\right\}$$

$$= \frac{2t^2}{n},$$

where we have used the facts that

$$E\left\{\Delta_j \mid \mathcal{F}_{t_j}\right\} = 0 \quad \text{and} \quad E\left\{(\Delta_j)^2\right\} = \frac{2t^2}{n^2}.$$

The latter is true because the fourth moment

$$E\left\{\left(B(t_{j+1}) - B(t_j)\right)^4\right\} = \frac{3t^2}{n^2}$$

as may be verified by differentiating the characteristic function of a normally distributed r.v. with variance t/n. By (4.2), $\lim_{n \to \infty} S_t^n = t$ in L^2. It then follows by Theorem 4.1(ii) that $[B]_t = t$ a.s. ∎

Proof of Theorem 4.1.

Proof of (i). Suppose M is bounded. Then M is a martingale by Proposition 1.8. We have

$$S_t^n = \sum_j \left\{(M_{t_{j+1}})^2 - (M_{t_j})^2 - 2M_{t_j}(M_{t_{j+1}} - M_{t_j})\right\}$$

(4.3)

$$= (M_t)^2 - (M_0)^2 - 2\int_0^t X^n \, dM.$$

Here X^n is a bounded predictable process defined for $(s, \omega) \in \mathbb{R}_+ \times \Omega$ by

$$X^n(s, \omega) = \sum_j M_{t_j}(\omega) 1_{(t_j, t_{j+1}]}(s).$$

By the continuity of M, $\lim_{n \to \infty} 1_{(0,t]} X^n = 1_{(0,t]} M$ pointwise on $\mathbb{R}_+ \times \Omega$ and hence in $\mathcal{L}^2$ by bounded convergence. Consequently,

$$\lim_{n \to \infty} \int_0^t X^n \, dM = \int_0^t M \, dM \quad \text{in } L^2,$$

by the isometry (equation (2.10)). The desired result follows from this and (4.3).

Proof of (ii). We now turn to the case where M is any continuous local martingale. Firstly we observe that by (2.27), the right member of (4.1) equals

$$(M_t)^2 - (M_0)^2 - 2M_0(M_t - M_0) - 2 \int_0^t (M - M_0) \, d(M - M_0)$$

$$= (M_t - M_0)^2 - 2 \int_0^t (M - M_0) \, d(M - M_0).$$

Since the value of this expression and of S_t^n remains unchanged when we replace M by $M - M_0$, we may and do suppose $M_0 = 0$. For each $k \in \mathbb{N}$, let $\tau_k = \inf\{s > 0 : |M_s| > k\}$. Then $\{\tau_k\}$ is a localizing sequence for M such that $M^k = \{M_{s \wedge \tau_k}, s \in \mathbb{R}_+\}$ is a bounded continuous martingale for each k. Since $M_s = M_s^k$ whenever $0 \le s \le t \le \tau_k$, then

$$1_{\{t \le \tau_k\}} S_t^n = 1_{\{t \le \tau_k\}} \sum_j \left(M_{t_{j+1}}^k - M_{t_j}^k \right)^2.$$

By applying part (i) to M^k, it follows that the sum on the right above converges in L^2 and therefore in pr. to $[M^k]_t$ as $n \to \infty$. On $\{\omega : t \le \tau_k(\omega)\}$, we have $M^k = M$ and $\int_0^t M^k \, dM^k = \int_0^t M \, dM$ a.s., by the definition of the latter (see the paragraph following Lemma 2.10). Thus by (4.1) we have

$$1_{\{t \le \tau_k\}} [M^k]_t = 1_{\{t \le \tau_k\}} [M]_t \text{ a.s.}$$

Therefore, for each k, as $n \to \infty$:

$$1_{\{t \le \tau_k\}} S_t^n \to 1_{\{t \le \tau_k\}} [M]_t \text{ in pr.}$$

Since $P\left(\bigcup_k \{t \le \tau_k\}\right) = 1$, it follows that $S_t^n \to [M]_t$ in pr. ∎

4.3 Properties of Quadratic Variation for an L^2-Martingale

In the next theorem, we list some results for the case where M is a continuous L^2-martingale, and give a basic representation for the Doléans measure μ_M in terms of $[M]$.

Definition. A process $U = \{U_t, t \in \mathbb{R}_+\}$ will be called:

(i)   *increasing* iff U is adapted and almost surely the sample function $t \to U_t(\omega)$ is increasing on $\mathbb{R}_+$;

(ii)   *integrable* iff $U_t \in L^1$ for each t.

Theorem 4.2. *Let M be a continuous L^2-martingale. Then the following hold.*

(i) $[M] = \{[M]_t, t \in \mathbb{R}_+\}$ *is a continuous integrable increasing process with $[M]_0 = 0$.*

(ii) $\{\int_0^t M \, dM, t \in \mathbb{R}_+\}$ *is a continuous martingale with zero mean.*

(iii)   *For each t, the sequence $\{S_t^n, n \in \mathbb{N}\}$, defined in Theorem 4.1, converges in L^1 to $[M]_t$.*

(iv)   *The content $\lambda_{[M]}$ of $[M]$, defined in Chapter 2, is given by*

(4.4)
$$\lambda_{[M]}(A) = E\left(\int_0^\infty 1_A(s) \, d[M]_s\right)$$

for each $A \in \mathcal{A}$, and furthermore $\lambda_{[M]} = \mu_M$ on $\mathcal{A}$.

(v)   *For $X \in \Lambda^2(\mathcal{P}, M)$ and each t,*

(4.5)
$$E\left(\int_0^t (X_s)^2 \, d[M]_s\right) = \int_{\mathbb{R}_+ \times \Omega} 1_{[0,t]}(X)^2 \, d\mu_M.$$

(vi)   *For $X \in \mathcal{L}^2(\mathcal{P}, \mu_M)$,*

$$E\left(\int_0^\infty (X_s)^2 \, d[M]_s\right) = \int_{\mathbb{R}_+ \times \Omega} (X)^2 \, d\mu_M.$$

Proof. For part (i), the continuity and adaptedness of $[M]$ and the fact that $[M]_0 = 0$ follow from (4.1). To prove integrability, for fixed t let S_t^n be defined as in Theorem 4.1. By the L^2-integrability of M, the orthogonality property: $E\{M_r(M_s - M_r)\} = 0$ for $r < s$, and the first equality in (4.3) which holds even when M is not bounded, it follows that $S_t^n \in L^1$ and

(4.6) $$E(S_t^n) = E\{(M_t)^2 - (M_0)^2\}.$$

As $n \to \infty$, $S_t^n \to [M]_t$ in pr. Hence by Fatou's lemma we have

$$E([M]_t) \leq E\{(M_t)^2 - (M_0)^2\}.$$

Thus $[M]_t \in L^1$, proving $[M]$ is integrable. To prove $t \to [M]_t(\omega)$ is increasing for almost every ω, it suffices to show for fixed $s < t$ that $[M]_s \leq [M]_t$ a.s., since $[M]$ has continuous paths. But this follows from Theorem 4.1(ii). For if $\{\pi_s^n, n \in I\!N\}$ is a sequence of partitions of $[0, s]$ with $\delta\pi_s^n \to 0$, this can be extended by the addition of points in $(s, t]$ to a sequence of partitions $\{\pi_t^n, n \in I\!N\}$ of $[0, t]$ such that $\delta\pi_t^n = \delta\pi_s^n$ and $S_s^n \leq S_t^n$ for each n. Applying Theorem 4.1(ii) and letting $n \to \infty$ through a subsequence of n's leads to the desired result: $[M]_s \leq [M]_t$ a.s.

For the proof of part (ii), by Theorem 2.11, $\{\int_0^t M \, dM, t \in I\!R_+\}$ is a continuous local martingale. Let $\{\tau_k\}$ be a localizing sequence for it. To prove it is a martingale, it suffices by Proposition 1.8 to show that $\left\{\left|\int_0^{t \wedge \tau_k} M \, dM\right|, k \in I\!N\right\}$ is uniformly integrable for each t. By (4.1) we have

$$\int\limits_0^{t \wedge \tau_k} M \, dM = \frac{1}{2} \left\{ (M_{t \wedge \tau_k})^2 - (M_0)^2 - [M]_{t \wedge \tau_k} \right\}.$$

Since $[M]$ is a.s. increasing, it follows that

$$(4.7) \qquad \left| \int\limits_0^{t \wedge \tau_k} M \, dM \right| \leq \frac{1}{2} \left\{ (M_{t \wedge \tau_k})^2 + (M_0)^2 + [M]_t \right\} \qquad \text{a.s.}$$

By Corollary 1.7 to Doob's Stopping Theorem, since the sequence of optional times $\{t \wedge \tau_k, k \in \mathbb{N}\}$ is bounded by t, $\{(M_{t \wedge \tau_k})^2, k \in \mathbb{N}\}$ is uniformly integrable. Also, $(M_0)^2$ and $[M]_t$ are in L^1. It follows that $\{\left| \int_0^{t \wedge \tau_k} M \, dM \right|, k \in \mathbb{N}\}$ is uniformly integrable as required. Hence $\{\int_0^t M \, dM, t \in \mathbb{R}_+\}$ is a martingale and its mean equals that of its initial r.v., which is zero.

By the zero mean property just proved, and (4.1) and (4.6), we have $E(S_t^n) = E([M]_t)$ for each n and t. Since S_t^n and $[M]_t$ are positive r.v.'s, it follows by Proposition 1.1 that $\{S_t^n, n \in \mathbb{N}\}$ converges in L^1 to $[M]_t$ for each t. Thus, part (iii) has been proved.

For the proof of part (iv), it suffices by linearity to prove the following:

$$(4.8) \qquad \lambda_{[M]}(R) = E\left(\int\limits_0^\infty 1_R \, d[M]_s \right) = \mu_M(R)$$

for all predictable rectangles R. When $R = 1_{\{0\} \times F_0}$ for $F_0 \in \mathcal{F}_0$, then $\lambda_{[M]}(R)$ and $\mu_M(R)$ are zero by definition, and $\int_0^\infty 1_R \, d[M]_s = 0$ since $s \rightarrow [M]_s$ is continuous at $s = 0$. When $R = 1_{(s,t] \times F}$, for $s < t$ and $F \in \mathcal{F}_s$,

$$\lambda_{[M]}(R) = E\{1_F([M]_t - [M]_s)\} = E\left(\int\limits_0^\infty 1_R \, d[M]_s \right)$$

and

$$\mu_M(R) = E\left\{1_F\left((M_t)^2 - (M_s)^2\right)\right\}$$

$$= E\left\{1_F\left([M]_t - [M]_s + 2\int_s^t M\,dM\right)\right\}.$$

The last equality above is by (4.1). Now by part (ii) we have

$$E\left(\int_s^t M\,dM \,\Big|\, \mathcal{F}_s\right) = E\left(\int_0^t M\,dM - \int_0^s M\,dM \,\Big|\, \mathcal{F}_s\right) = 0.$$

It follows that $E\left\{1_F\left(\int_s^t M\,dM\right)\right\} = 0$. By combining the above, we conclude that (4.8) holds and hence so does part (iv).

The right side of (4.4) defines a measure on sets A in $\mathcal{P}$, which we shall also denote by $\lambda_{[M]}$. By part (iv) above, $\lambda_{[M]}$ agrees with μ_M on $\mathcal{A}$, hence on $\mathcal{P}$, by the uniqueness of the extension from $\mathcal{A}$ to $\mathcal{P}$. Let $t \in \mathbb{R}_+$. If X is the indicator function of a set in $\mathcal{P}$, then $1_{[0,t]}(X)^2$ is also of this form, and hence by the equality of $\lambda_{[M]}$ and μ_M on $\mathcal{P}$, (4.5) holds in this case. Then by linearity, (4.5) holds whenever X is a $\mathcal{P}$-simple function. For general $X \in \Lambda^2(\mathcal{P}, M)$, there is a sequence $\{X^n\}$ of $\mathcal{P}$-simple functions such that $1_{[0,t]}(X^n)^2 \uparrow 1_{[0,t]}(X)^2$ as $n \to \infty$ and so (4.5) follows for X by monotone convergence.

Property (vi) follows by monotone convergence on letting $t \to \infty$ in (v). ∎

4.4 Direct Definition of μ_M

Our definition (4.1) of $[M]$ and proof of Theorem 4.1 used the stochastic integral and consequently the result, stated without proof in Chapter 2, that μ_M can be extended from $\mathcal{A}$ to a measure on $\mathcal{P}$. When M is a continuous bounded martingale, we can prove directly that a sequence of quadratic sums $\{S_t^n, n \in \mathbb{N}\}$ converges in L^2 to a limit S_t such that $S = \{S_t, t \in \mathbb{R}_+\}$ is a continuous increasing process and $\{(M_t)^2 - (M_0)^2 - S_t, t \in \mathbb{R}_+\}$ is a martingale. Then the right member of (4.4), with $[M]$ there replaced by

S, defines a measure on sets A in $\mathcal{P}$ which agrees with μ_M on $\mathcal{A}$, since S has all the properties of $[M]$ required for the proof of Theorem 4.2(iv). Consequently, for such M, μ_M can be defined directly on $\mathcal{P}$. This approach and the following theorem, in which the convergence of $\{S_t^n, n \in I\!N\}$ is proved, is taken from notes by M. J. Sharpe.

Theorem 4.3. *Let M be a continuous bounded martingale. For each t and n let*

$$t_j^n = t \wedge (j2^{-n}) \text{ for } j \in I\!N_0, \quad \text{and}$$

$$S_t^n = \sum_{j=0}^{\infty} \left(M_{t_{j+1}^n} - M_{t_j^n} \right)^2.$$

Then, for each t, the sequence $\{S_t^n, n \in I\!N\}$ converges in L^2. If S_t denotes the L^2-limit of this sequence, then $\{(M_t)^2 - (M_0)^2 - S_t, t \in I\!R_+\}$ is a martingale and there is a version of $\{S_t, t \in I\!R_+\}$ which is a continuous increasing process.

Proof. Suppressing the superscripts n on the t_j^n, we rewrite S_t^n as follows:

(4.9)
$$S_t^n = \sum_{j=0}^{\infty} \left\{ \left(M_{t_{j+1}} \right)^2 - \left(M_{t_j} \right)^2 - 2 M_{t_j} \left(M_{t_{j+1}} - M_{t_j} \right) \right\}$$

$$= (M_t)^2 - (M_0)^2 - 2 \sum_{j=0}^{\infty} M_{t_j} \left(M_{t_{j+1}} - M_{t_j} \right).$$

For each fixed n and t, in any bounded interval the sum in the last line of (4.9) consists of only finitely many non-zero terms, each of which is a continuous martingale. It follows that the sum is a continuous martingale for each n.

Let t be fixed. To prove that $\{S_t^n, n \in I\!N\}$ converges in L^2, we show it is a Cauchy sequence in L^2. For $m < n$ fixed, let $t_j' = t \wedge (2^{-m}[j2^{m-n}])$ for each $j \in I\!N_0$ where $[\,\cdot\,]$ denotes "integer part". Then $t_j' \le t_j, |t_j - t_j'| \le 2^{-m}$ and by (4.9),

$$S_t^n - S_t^m = -2 \sum_{j=0}^{[2^n t]} \left(M_{t_j} - M_{t_j'} \right) \left(M_{t_{j+1}} - M_{t_j} \right).$$

For each j, let $Z_j = M_{t_j} - M_{t'_j}$ and note that $Z_j \in \mathcal{F}_{t_j}$. Then for $j < k$, $Z_j \left(M_{t_{j+1}} - M_{t_j} \right) Z_k \in \mathcal{F}_{t_k}$ is orthogonal to $M_{t_{k+1}} - M_{t_k}$ in the sense that

$$(4.10) \qquad E \left\{ Z_j \left(M_{t_{j+1}} - M_{t_j} \right) Z_k \left(M_{t_{k+1}} - M_{t_k} \right) \right\} = 0.$$

Hence,

$$(4.11) \qquad E \left\{ (S_t^n - S_t^m)^2 \right\} = 4 E \left\{ \sum_{j=0}^{[2^n t]} Z_j^2 \left(M_{t_{j+1}} - M_{t_j} \right)^2 \right\},$$

since the cross-product terms are zero by (4.10). Let

$$W_t^m = \sup_{r,s} (M_s - M_r)^2,$$

where the supremum is over all $0 \le r \le s \le t$ such that $0 \le s - r \le 2^{-m}$. Then the expectation on the right side of (4.11) is dominated by

$$E \left\{ W_t^m \sum_{j=0}^{[2^n t]} \left(M_{t_{j+1}} - M_{t_j} \right)^2 \right\}.$$

By Schwarz's inequality the above is less than or equal to

$$(4.12) \qquad \left[E \left\{ (W_t^m)^2 \right\} E \left\{ \left(\sum_{j=0}^{[2^n t]} \left(M_{t_{j+1}} - M_{t_j} \right)^2 \right)^2 \right\} \right]^{\frac{1}{2}}.$$

By the uniform continuity of $u \to M_u(\omega)$ on $[0, t]$, we have

$$\lim_{m \to \infty} W_t^m(\omega) = 0.$$

Since M is bounded, it follows that this convergence is also in L^2. To show that the second expectation in (4.12) is bounded, we let $C > 0$ be a bound for M and $a_j = \left(M_{t_{j+1}} - M_{t_j} \right)^2$ for each j. Then the expectation is given by

$$E \left\{ \left(\sum_{j=0}^{[2^n t]} a_j \right)^2 \right\} = E \left\{ \sum_j a_j^2 + 2 \sum_j a_j \sum_{k>j} a_k \right\}$$

$$= E \left\{ \sum_j a_j^2 \right\} + 2 E \left\{ \sum_j a_j \, E \left(\sum_{k>j} a_k \Big| \mathcal{F}_{t_{j+1}} \right) \right\}$$

In the above,

$$\sum_j a_j^2 \le \left(\max_l a_l \right) \left(\sum_j a_j \right) \le 4C^2 \sum_j a_j$$

and by (2.3) we have

$$
\begin{aligned}
E\left(\sum_{k>j} a_k \,\Big|\, \mathcal{F}_{t_{j+1}} \right) &= \sum_{k>j} E\left\{ \left(M_{t_{k+1}} - M_{t_k} \right)^2 \,\Big|\, \mathcal{F}_{t_{j+1}} \right\} \\
&= \sum_{k>j} E\left\{ \left(M_{t_{k+1}} \right)^2 - \left(M_{t_k} \right)^2 \,\Big|\, \mathcal{F}_{t_{j+1}} \right\} \\
&= E\left\{ \left(M_t \right)^2 - \left(M_{t_{j+1}} \right)^2 \,\Big|\, \mathcal{F}_{t_{j+1}} \right\} \\
&\le C^2.
\end{aligned}
$$

By combining the above and using (4.6) we obtain

$$
\begin{aligned}
E\left\{ \left(\sum_{j=0}^{[2^n t]} a_j \right)^2 \right\} &\le (4C^2 + 2C^2)\, E\left(\sum_j a_j \right) \\
&= 6C^2 E\left\{ \left(M_t \right)^2 - \left(M_0 \right)^2 \right\}.
\end{aligned}
$$

It follows that as n and $m \to \infty$, (4.12) tends to zero, and hence $S_t^n - S_t^m \to 0$ in L^2. Thus $\{S_t^n, n \in I\!N\}$ is a Cauchy sequence and hence converges in L^2.

For each t, let S_t denote the L^2-limit of $\{S_t^n, n \in I\!N\}$. Since the martingale property is preserved for L^2-limits, it follows from (4.9) that $Y = \{(M_t)^2 - (M_0)^2 - S_t, t \in I\!R_+\}$ is an L^2-martingale. By the same reasoning as used in the proof of Theorem 2.5, there is a continuous version of Y which we shall again denote by Y. Then $\{(M_t)^2 - (M_0)^2 - Y_t, t \in I\!R_+\}$ is a continuous version of S, and since S_t is only defined as an L^2-limit, we may let $S_t \equiv (M_t)^2 - (M_0)^2 - Y_t$ for all t. Since M and Y are adapted, so is S. To prove that S is increasing, it suffices by path continuity to show for $s < t$ that $S_s \le S_t$ a.s. Now,

$$S_t - S_s = \lim_{n \to \infty} \left(S_t^n - S_s^n \right) \quad \text{in } L^2.$$

Using the same t_j notation as earlier and letting $k = \inf \{j : t_j > s\}$, we have

$$S_t^n - S_s^n = \sum_{t_j > s} \left(M_{t_{j+1}} - M_{t_j}\right)^2 + \left(M_{t_k} - M_{t_{k-1}}\right)^2 - \left(M_s - M_{t_{k-1}}\right)^2.$$

As shown previously,

$$\left| \left(M_{t_k} - M_{t_{k-1}}\right)^2 - \left(M_s - M_{t_{k-1}}\right)^2 \right| \le 2W_t^n \to 0 \text{ in } L^2$$

as $n \to \infty$. Thus $S_t - S_s = \lim_{n \to \infty} \sum_{t_j > s} \left(M_{t_{j+1}} - M_{t_j}\right)^2$ in L^2, and hence is positive a.s. ∎

When M is a continuous L^2-martingale, we can also prove the extendibility of μ_M from $\mathcal{A}$ to $\mathcal{P}$ as follows. We may assume $M_0 = 0$ since M and $M - M_0$ induce the same content μ_M on $\mathcal{A}$. Since M is a local martingale, there is a localizing sequence $\{\tau_k\}$ such that for each k, $M^k = \{M_{t \wedge \tau_k}, t \in \mathbb{R}_+\}$ is bounded, and thus μ_{M^k} can be defined on $\mathcal{P}$ as above. Then $\int_0^t M \, dM = \lim_{k \to \infty} \int_0^t M^k \, dM^k$ defines a continuous local martingale without the use of μ_M. Equation (4.1), Theorems 4.1 and 4.2 then follow. In particular, the right side of (4.4) defines a measure on $\mathcal{P}$ which agrees with μ_M on $\mathcal{A}$.

4.5 Decomposition of $(M)^2$

Next we shall discuss the decomposition afforded by (4.1) of the continuous local submartingale $(M_t)^2$ as the sum of the continuous local martingale $(M_0)^2 + 2 \int_0^t M \, dM$ and the continuous increasing process $[M]_t$. We begin with the canonical example of Brownian motion.

Example. If B is a Brownian motion in $\mathbb{R}$ with $B_0 \in L^2$, then by (4.1) and the example following it we have a.s.

$$(B_t)^2 = (B_0)^2 + 2 \int_0^t B \, dB + t.$$

By Theorem 4.2, $\{(B_0)^2 + 2 \int_0^t B \, dB, t \in \mathbb{R}_+\}$ is a continuous martingale.

Clearly, $\{t, t \in \mathbb{R}_+\}$ is a continuous integrable increasing process with initial value zero.

For a continuous L^2-martingale, M, by (4.1) we have the following decomposition of the continuous submartingale $(M)^2$:

$$(4.13) \qquad (M_t)^2 = (M_0)^2 + 2 \int_0^t M \, dM + [M]_t.$$

By Theorem 4.2, $\{(M_0)^2 + 2 \int_0^t M \, dM, t \in \mathbb{R}_+\}$ is a continuous martingale and $[M]$ is a continuous integrable increasing process with initial value zero. This decomposition is in fact unique (up to indistinguishability) and is a special case of the Doob-Meyer decomposition theorem (see Dellacherie and Meyer [24, VII-9]). The proof we give of the uniqueness depends on Lemma 4.4, for which we need the following.

Definition. A process V is said to be *locally of bounded variation* iff it is adapted and almost surely the sample function $t \rightarrow V_t(\omega)$ is of bounded variation on each bounded interval in $\mathbb{R}_+$.

Remark. Some authors say a process V is locally of bounded variation if it is adapted and there is a sequence of optional times $\{\tau_k\}$ increasing to ∞ such that almost surely the sample function $t \rightarrow V_{t \wedge \tau_k}(\omega)$ is of bounded variation on $\mathbb{R}_+$ for each k. This is equivalent to our definition. We only treat the case of continuous V here. A process is continuous and locally of bounded variation if and only if it is the difference of two continuous increasing processes.

Lemma 4.4. *Let V be a continuous L^2-martingale that is locally of bounded variation. Then*

$$P(V_t = V_0 \quad \text{for all } t \in \mathbb{R}_+) = 1.$$

Proof. Let $\hat{V}(t) = V(t) - V(0)$. By the path continuity of V, it suffices to prove $P(\hat{V}(t) = 0) = 1$ for each t. Since $\hat{V}$ is continuous and locally of bounded variation, by ordinary calculus applied path-by-path, we have for each t,

$$(\hat{V}_t)^2 = (\hat{V}_t)^2 - (\hat{V}_0)^2 = 2 \int_0^t \hat{V}_s \, d\hat{V}_s.$$

The integral here is the same (a.s.) whether viewed as a path-by-path Riemann-Stieltjes integral or a stochastic integral. Viewing it as the latter, by Theorem 4.2(ii) it defines a continuous martingale and hence

$$E \left[2 \int_0^t \hat{V}_s \, d\hat{V}_s \right] = 0.$$

Combining the above, we obtain $E[(\hat{V}_t)^2] = 0$ and hence $\hat{V}_t = 0$ a.s., as desired. ∎

Corollary 4.5. *Let V be a continuous local martingale that is locally of bounded variation. Then*

$$P(V_t = V_0 \quad \text{for all } t \in \mathbb{R}_+) = 1.$$

Proof. Apply Lemma 4.4 to appropriate localizations $\hat{V}^k$ of $V - V_0$ and then let $k \to \infty$. ∎

The following decomposition theorem is an immediate consequence of (4.13), Lemma 4.4 and the fact that the difference of two continuous increasing processes is continuous and locally of bounded variation.

Theorem 4.6. *Let M be a continuous L^2-martingale. Then there is a a unique decomposition of $(M)^2$ as the sum of a continuous martingale and a continuous integrable increasing process with initial value zero. This decomposition is given by*

$$(4.14) \qquad (M_t)^2 = \left((M_0)^2 + 2 \int_0^t M \, dM \right) + [M]_t \quad \text{for all } t \geq 0.$$

If M is a continuous *local* martingale, equation (4.14) still holds, but then $(M_0)^2 + 2 \int_0^t M \, dM$ is a *local* martingale and $[M]$ need not be integrable. The uniqueness of the decomposition follows from Corollary 4.5. Stated formally we have the following.

Theorem 4.7. *Let M be a continuous local martingale. Then there is a unique decomposition of $(M)^2$ as the sum of a continuous local martingale and a continuous increasing process with initial value zero. This decomposition is given by (4.14).*

4.6 A Limit Theorem

The following theorem will be needed for the proof of the Itô formula in the next chapter.

Theorem 4.8. *Let M be a continuous local martingale and let Y be a bounded continuous adapted process. Let $t \in I\!R_+$ and $\{\pi_t^n, n \in I\!N\}$ be a sequence of partitions of $[0, t]$ such that $\lim_{n \to \infty} \delta \pi_t^n = 0$. For each $n \in I\!N$ let*

$$Z_n = \sum_j Y_{t_j} \left(M_{t_{j+1}} - M_{t_j} \right)^2$$

where the sum is over all j such that t_j and t_{j+1} are both in π_t^n. Then $\{Z_n, n \in I\!N\}$ converges in probability to $\int_0^t Y_s \, d[M]_s$.

Proof. For each $n \in I\!N$, define

$$W_n = \sum_j Y_{t_j} \left([M]_{t_{j+1}} - [M]_{t_j} \right),$$

where the sum is over all j such that $t_j, t_{j+1} \in \pi_t^n$. The sequence $\{W_n, n \in I\!N\}$ converges a.s. and hence in pr. to $\int_0^t Y_s \, d[M]_s$. To prove that $\{Z_n\}$ converges in pr. to the same limit, it suffices to show that R_n, defined by

(4.15)
$$\begin{aligned} R_n &\equiv Z_n - W_n \\ &= \sum_j Y_{t_j} \left\{ \left(M_{t_{j+1}} - M_{t_j} \right)^2 - \left([M]_{t_{j+1}} - [M]_{t_j} \right) \right\}. \end{aligned}$$

converges to zero in pr. As in the proof of Theorem 4.1(ii), it suffices to consider the case where M is bounded. Then for $0 \le r < s \le t$, we have the following

$$(M_s - M_r)^2 - ([M]_s - [M]_r)$$
$$= ((M_s)^2 - [M]_s) - ((M_r)^2 - [M]_r) - 2M_r(M_s - M_r)$$

(4.16)

$$= 2\left\{ \int_r^s M\, dM - M_r(M_s - M_r) \right\}$$

The second equality above follows by (4.1). For each n, let Y^n and M^n be the processes defined by

$$Y^n = \sum_j Y_{t_j} 1_{(t_j, t_{j+1}]} \quad \text{and} \quad M^n = \sum_j M_{t_j} 1_{(t_j, t_{j+1}]}.$$

Then by (4.15) and (4.16), R_n is equal to

$$2\sum_j Y_{t_j} \left\{ \int_{t_j}^{t_{j+1}} M\, dM - M_{t_j}\left(M_{t_{j+1}} - M_{t_j}\right) \right\}$$
$$= 2\sum_j \left\{ \int 1_{(t_j, t_{j+1}]} Y_{t_j} M\, dM - \int 1_{(t_j, t_{j+1}]} Y_{t_j} M_{t_j}\, dM \right\}$$
$$= 2\int_0^t Y^n(M - M^n)\, dM.$$

The first equality above is by Theorem 2.7(i). By the continuity of Y and M, $1_{[0,t]} Y^n(M - M^n)$ converges to zero pointwise on $\mathbb{R}_+ \times \Omega$ as $n \to \infty$, and hence in $\mathcal{L}^2$ by bounded convergence. Consequently, by the isometry, R_n converges to zero in L^2 and hence in pr. ∎

4.7 Exercises

1. Suppose M is a continuous L^2-bounded martingale. Then by the martingale convergence theorem (Theorem 1.5), $\lim_{t \to \infty} M_t$ exists a.s. and defines a random variable $M_\infty \in L^2$. Moreover, since $[M]$ is an increasing process, P-a.s., $\lim_{t \to \infty}[M]_t$ exists in $\overline{\mathbb{R}}_+$. Let $[M]_\infty$ denote the extended

real-valued random variable defined by this limit. Show that

$$E([M]_\infty) = E(M_\infty^2 - M_0^2) = E((M_\infty - M_0)^2) < \infty.$$

Deduce that $[M]_\infty < \infty$ a.s. and $\lim_{t\to\infty} \int_0^t M \, dM$ exists a.s. and defines a random variable in L^1.

2.   Suppose M is a continuous local martingale and let $[M]_\infty = \lim_{t\to\infty}[M]_t$. Show that if $E([M]_\infty) < \infty$ then $M - M_0$ is an L^2-bounded martingale.

3.   Suppose M is a continuous local martingale and V is a continuous process that is locally of bounded variation. Define $X = M + V$. Then X is called a *semimartingale*. Indeed, a continuous semimartingale is any continuous adapted process that can be decomposed in this way. Such a decomposition is unique if we require that $V_0 = 0$ P-a.s. Let $t \in I\!\!R_+$ and $\{\pi_t^n, n \in I\!\!N\}$ be a sequence of partitions of $[0, t]$ as in Section 4.2. For each n, let

$$S_t^n = \sum_j (X_{t_{j+1}} - X_{t_j})^2,$$

where the sum is over j such that t_j and t_{j+1} are both in π_t^n. Prove that $\{S_t^n, n \in I\!\!N\}$ converges in probability to $[M]_t$.

4.   Suppose that M is a continuous, bounded martingale and let S be defined as in Section 4.4. Verify that the right member of (4.4), with S in place of $[M]$ there, defines a measure on the predictable σ-field $\mathcal{P}$.

Hint: Prove this first for predictable sets A in $[0, t] \times \Omega$ using a monotone class theorem.

5

THE ITO FORMULA

5.1 Introduction

One of the most important results in the theory of stochastic integrals is the rule for change of variables known as the Itô formula, after Itô who first proved it for the special case of integration with respect to Brownian motion. The essential aspects of Itô's formula are conveyed by the following. If M is a continuous local martingale and f is a twice continuously differentiable real-valued function on $\mathbb{R}$, then the Itô formula for $f(M_t)$ is

$$(5.1) \qquad f(M_t) - f(M_0) = \int_0^t f'(M_s)\, dM_s + \frac{1}{2} \int_0^t f''(M_s)\, d[M]_s.$$

Compare this with the fundamental theorem of calculus for real variables. Here there is an additional term involving the quadratic variation process.

When $f(x) = x^2$, (5.1) reduces to the definition of the quadratic variation process.

5.2 One-Dimensional Itô Formula

Consider a pair (M, V) where M is a continuous local martingale and V is a continuous process which is locally of bounded variation. The Itô formula for this pair is stated below. Since M and V are real-valued processes, this is often referred to as the one-dimensional Itô formula. The multi-dimensional Itô formula, for vector-valued processes, is discussed later in this chapter.

Theorem 5.1. *Let M be a continuous local martingale and V be a continuous process which is locally of bounded variation. Let f be a continuous real-valued function defined on $I\!R^2$ such that the partial derivatives $\frac{\partial f}{\partial x}(x, y)$, $\frac{\partial^2 f}{\partial x^2}(x, y)$, and $\frac{\partial f}{\partial y}(x, y)$, exist and are continuous for all (x, y) in $I\!R^2$. Then a.s., we have for each t*

$$
f(M_t, V_t) - f(M_0, V_0) = \int_0^t \frac{\partial f}{\partial x}(M_s, V_s)\, dM_s
$$

(5.2)
$$
+ \int_0^t \frac{\partial f}{\partial y}(M_s, V_s)\, dV_s
$$

$$
+ \frac{1}{2} \int_0^t \frac{\partial^2 f}{\partial x^2}(M_s, V_s)\, d[M]_s.
$$

For clarity, we have put in the time-parameter s in the stochastic integral $\int_0^t \frac{\partial f}{\partial x}(M_s, V_s)\, dM_s$. The reader should keep in mind that this integral is defined stochastically, not path-by-path. A suggestive way to write (5.2) is by using differentials:

(5.3)
$$
df(M_t, V_t) = \frac{\partial f}{\partial x}(M_t, V_t) dM_t
$$
$$
+ \frac{\partial f}{\partial y}(M_t, V_t) dV_t + \frac{1}{2} \frac{\partial^2 f}{\partial x^2}(M_t, V_t) d[M]_t.
$$

Of course the rigorous interpretation of (5.3) is the integrated form (5.2).

Example. Let B denote a Brownian motion in $I\!R$ and let $f(x) = x^2$. With $M = B$ and $f(x, y) = f(x)$, (5.3) becomes

$$d(B_t)^2 = 2B_t dB_t + dt.$$

Formally this suggests $(dB_t)^2 = dt$. For general M the appropriate formalism is $(dM_t)^2 = d[M]_t$. Heuristically this explains the presence of the additional term in the Itô formula.

Proof of Theorem 5.1. Since both sides of the equality in (5.2) are continuous processes, it suffices to prove for each t that (5.2) holds a.s. Let $\{\pi_t^n, n \in I\!N\}$ be a sequence of partitions of $[0, t]$ such that $\lim_{n \to \infty} \delta \pi_t^n = 0$. We use the same notation for members of π_t^n as in Chapter 4. In particular, we omit the superscript n from t_j^n. Thus,

(5.4)
$$
f(M_t, V_t) - f(M_0, V_0) = \sum_j \left\{ f\left(M_{t_{j+1}}, V_{t_{j+1}}\right) \right.
$$
$$
- f\left(M_{t_{j+1}}, V_{t_j}\right)
$$
$$
\left. + f\left(M_{t_{j+1}}, V_{t_j}\right) - f\left(M_{t_j}, V_{t_j}\right) \right\}.
$$

By Taylor's theorem, the right side of the equals sign above may be written as:

(5.5)
$$
\sum_j \left\{ \left(\left(\frac{\partial f}{\partial y} \right) (M_{t_j}, V_{t_j}) + \varepsilon_j^1 \right) \left(V_{t_{j+1}} - V_{t_j} \right) \right.
$$
$$
+ \frac{\partial f}{\partial x} (M_{t_j}, V_{t_j}) \left(M_{t_{j+1}} - M_{t_j} \right)
$$
$$
\left. + \frac{1}{2} \left(\frac{\partial^2 f}{\partial x^2} (M_{t_j}, V_{t_j}) + \varepsilon_j^2 \right) \left(M_{t_{j+1}} - M_{t_j} \right)^2 \right\}.
$$

where

$$
\varepsilon_j^1 = \frac{\partial f}{\partial y} \left(M_{t_{j+1}}, V_{\tau_j} \right) - \frac{\partial f}{\partial y} \left(M_{t_j}, V_{t_j} \right)
$$

and

$$
\varepsilon_j^2 = \frac{\partial^2 f}{\partial x^2} \left(M_{\eta_j}, V_{t_j} \right) - \frac{\partial^2 f}{\partial x^2} \left(M_{t_j}, V_{t_j} \right),
$$

for some random times τ_j and η_j in $[t_j, t_{j+1}]$. For each ω, the functions $(r,s) \to \frac{\partial f}{\partial y}(M_r, V_s)(\omega)$ and $(r,s) \to \frac{\partial^2 f}{\partial x^2}(M_r, V_s)(\omega)$ are uniformly continuous on $[0,t]^2$, and hence $\sup_j |\varepsilon_j^1(\omega)|$ and $\sup_j |\varepsilon_j^2(\omega)|$ tend to zero as $n \to \infty$. (Note that ε_j^1 and ε_j^2 depend on n although the notation does not specifically indicate this because the indices n on the t_j^n have been suppressed.) From this property of $\varepsilon_j^1(\omega)$, the continuity of $s \to \frac{\partial f}{\partial y}(M_s, V_s)(\omega)$, and since $s \to V_s(\omega)$ is of bounded variation on $[0,t]$ for almost every ω, it follows that

$$\sum_j \left(\frac{\partial f}{\partial y}(M_{t_j}, V_{t_j}) + \varepsilon_j^1 \right) (V_{t_{j+1}} - V_{t_j}) \to \int_0^t \frac{\partial f}{\partial y}(M_s, V_s)\, dV_s$$

almost surely as $n \to \infty$. From the above property of $\varepsilon_j^2(\omega)$, and since $\sum_j (M_{t_{j+1}} - M_{t_j})^2 \to [M]_t$ in pr. as $n \to \infty$ by Theorem 4.1(ii), it follows that $\sum_j \varepsilon_j^2 (M_{t_{j+1}} - M_{t_j})^2 \to 0$ in pr. as $n \to \infty$.

The proof will be completed in two steps. First we prove that when M and V are bounded, the terms in (5.5) involving the x-partial derivatives of f converge in pr. to the appropriate terms in (5.2). Then we extend (5.2) to the general case by using a localizing sequence for M and V.

Suppose that M and V are bounded. Then $\frac{\partial f}{\partial x}$ and $\frac{\partial^2 f}{\partial x^2}$ are bounded on the range of (M, V) and μ_M is a finite measure on $\mathcal{P}$. For each n, the process X^n defined by

$$X^n = \sum_j \frac{\partial f}{\partial x}(M_{t_j}, V_{t_j})\, 1_{(t_j, t_{j+1}]} + \frac{\partial f}{\partial x}(M_0, V_0)\, 1_{\{0\}}$$

is predictable and $\{X^n\}$ converges pointwise to $1_{[0,t]}\frac{\partial f}{\partial x}(M, V)$ on $\mathbb{R}_+ \times \Omega$ and hence by bounded convergence in $\mathcal{L}^2$. Then by the isometry,

$$\sum_j \frac{\partial f}{\partial x}(M_{t_j}, V_{t_j})(M_{t_{j+1}} - M_{t_j}) = \int X_s^n\, dM_s$$

$$\to \int_0^t \frac{\partial f}{\partial x}(M_s, V_s)\, dM_s$$

in L^2 as $n \to \infty$. Also, by Theorem 4.8 with $Y = \frac{\partial^2 f}{\partial x^2}(M, V)$, we have

$$\sum_j \frac{\partial^2 f}{\partial x^2}\left(M_{t_j}, V_{t_j}\right)\left(M_{t_{j+1}} - M_{t_j}\right)^2 \to \int_0^t \frac{\partial^2 f}{\partial x^2}(M_s, V_s)\, d[M]_s$$

in pr. as $n \to \infty$. It follows that the expression in (5.5) converges in pr. to the right side of (5.2) and hence (5.2) holds a.s. Thus we have proved the theorem when M and V are bounded.

To extend to the general case, for each n let $\tau_n = \inf\{t \geq 0 : |M_t| \vee |V_t| > n\}$. Then $M^n = M \cdot_{\wedge \tau_n} 1_{\{\tau_n > 0\}}$ and $V^n = V \cdot_{\wedge \tau_n} 1_{\{\tau_n > 0\}}$ are bounded, and it follows from the above that (5.2) holds a.s. with M^n, V^n in place of M, V, respectively, and hence holds a.s. with $t \wedge \tau_n$ in place of t. Then (5.2) follows on letting $n \to \infty$. ∎

Application.   A direct application of Theorem 5.1 yields the following representation for the stochastic integral $\int_0^t V_s\, dM_s$, where M is a continuous local martingale and V is a continuous process that is locally of bounded variation (see Exercise 2):

$$\int_0^t V_s\, dM_s = M_t V_t - M_0 V_0 - \int_0^t M_s\, dV_s.$$

Since the paths of M are continuous and V is locally of bounded variation, the integral on the right is defined pathwise as a Riemann-Stieltjes integral, and by integration by parts (cf. §1.3), the right member defines the Riemann-Stieltjes integral of V with respect to M. It follows that $\int_0^t V_s\, dM_s$ defines the same random variable whether the integral is regarded as a stochastic integral or a Riemann-Stieltjes integral.

Example. (*Ornstein-Uhlenbeck process.*) Let B denote a Brownian motion in $\mathbb{R}$ such that $B_0 = 0$ a.s. For $\alpha \in \mathbb{R}\backslash\{0\}$ and $\xi \in \mathcal{F}_0$, define

$$X_t = e^{-\alpha t}\xi + e^{-\alpha t}\int_0^t e^{\alpha s}\, dB_s, \quad \text{for all } t \geq 0.$$

We claim X is a solution of the stochastic differential equation:

$$dX_t = -\alpha X_t\, dt + dB_t.$$

This equation is known as *Langevin's equation*. It was originally introduced as a simple idealized model for the velocity of a particle suspended in a liquid. To be rigorously interpreted, this equation should be written in integrated form:

$$(5.6) \qquad\qquad X_t = X_0 - \alpha \int_0^t X_s\,ds + B_t.$$

To verify that X satisfies (5.6), let $M_t = \int_0^t e^{\alpha s}\,dB_s$ and $V_t = e^{-\alpha t}$. Then by the result of Exercise 2,

$$e^{-\alpha t}\int_0^t e^{\alpha s}\,dB_s = \int_0^t e^{-\alpha s}\,dM_s - \alpha \int_0^t M_s e^{-\alpha s}\,ds.$$

Then, using the substitution theorem 2.12 to expand dM_s in the first integral on the right, and substituting in the last integral for $M_s e^{-\alpha s}$ from the definition of X, we see that the right member above equals

$$B_t - \alpha \int_0^t X_s\,ds + \xi(1 - e^{-\alpha t}).$$

Substituting this for $e^{-\alpha t}\int_0^t e^{\alpha s}\,dB_s$ in the definition of X, we see that X satisfies (5.6).

Remarks.

1. Since the integral involved in the definition of X is the same whether it is considered as a stochastic integral or a Riemann-Stieltjes integral, one could have obtained X as a solution of (5.6) using the method of integrating factors for ordinary differential equations. On the other hand, when the coefficient of dB in a stochastic differential equation is not locally of bounded variation, the formal manipulations of ordinary Newton calculus need not apply. For instance, $(dB_t)^2 \neq 2B_t dB_t$. For another example, see Theorem 6.2, especially (6.7).

2. The process X defined above is known as the one-dimensional Ornstein-Uhlenbeck process with parameter α and initial value ξ. Using the path-by-path definition of $\int_0^t e^{\alpha s}\,dB_s$ as a Riemann-Stieltjes integral, we see that this integral is the almost sure limit of sums of the form $\sum_k \exp(\alpha k 2^{-n})\Delta_{k,t}B$ where the $\Delta_{k,t}B \equiv B\left(\frac{(k+1)}{2^n}\wedge t\right) - B\left(\frac{k}{2^n}\wedge t\right)$ are independent and normally

distributed random variables. It follows that $\int_0^t e^{\alpha s}\,dB_s$ defines a Gaussian process. Hence, if $X_0 = x \in I\!R$, then X is a Gaussian process with mean $E[X_t] = e^{-\alpha t}x$ and covariance $\Gamma(s,t) \equiv E[X_s X_t] - E[X_s]E[X_t] = (e^{-\alpha(t-s)} - e^{-\alpha(t+s)})/2\alpha$ for $s < t$. It can be shown that if $\alpha > 0$, X_t converges in distribution as $t \to \infty$ to a normal random variable with zero mean and variance $1/2\alpha$ (cf. Breiman [7; Chapter 16, Section 1]).

5.3 Mutual Variation Process

We now introduce the mutual variation process associated with two continuous local martingales. This notion is required for the multi-dimensional Itô formula.

Let M and N be continuous local martingales. Then so too are $M + N$ and $M - N$. Hence, by Theorem 4.7, $(M + N)^2 - [M + N]$ and $(M - N)^2 - [M - N]$ are local martingales and consequently so is

$$(5.7) \quad \begin{aligned} &MN - \frac{1}{4}\left([M + N] - [M - N]\right) \\ &= \frac{1}{4}\left((M + N)^2 - [M + N] - (M - N)^2 + [M - N]\right). \end{aligned}$$

Definition. For continuous local martingales M and N, let

$$[M, N] \equiv \frac{1}{4}\left([M + N] - [M - N]\right).$$

We call $[M, N]$ the *mutual variation process* of M and N.

Remark. Note that when $M = N$, $[M, M] = [M]$. Thus the mutual variation process is an extension of the quadratic variation process, just as covariance is an extension of variance. For this reason some authors refer to $[M, N]$ as the covariation process. Actually, in the general theory of *right* continuous local L^2-martingales there are two processes which may be so named. For a pair of right continuous local L^2-martingales, M and N, one of these, denoted by $[M, N]$, is defined as in the above. There the quadratic variation processes $[M+N]$ and $[M-N]$ are defined as limits in P-probability of sums of the form S_t^n as in Theorem 4.1, with $M + N$ and $M -$

N in place of M there. The other, denoted by $\langle M, N \rangle$, is defined to be the unique predictable process that is locally of bounded variation, starts from zero, and is such that $MN - \langle M, N \rangle$ is a local martingale. When M and N are continuous, $[M, N]$ is continuous, hence predictable, and so $\langle M, N \rangle = [M, N]$. Thus, in the situation we consider here, the two processes coincide and we will refer to this process as the mutual variation process and denote it by $[M, N]$. For further details on the general case, see Dellacherie and Meyer [24, VII.39–44]. Note that they adopt a slightly different convention regarding initial values and so their covariation processes differ from those described above by addition of the term $M_0 N_0$.

The process $[M, N]$ is the difference of two continuous increasing processes and hence is a continuous process which is locally of bounded variation. Therefore by (5.7) we have a decomposition of MN which is formally described by the following theorem.

Theorem 5.2. *Let M and N be continuous local martingales. Then there is a unique decomposition of MN as the sum of a continuous local martingale and a continuous process which is locally of bounded variation and has initial value zero. This decomposition is given for each t by:*

$$(MN)_t = (MN)_0 + \frac{1}{2} \int_0^t (M + N)_s \, d(M + N)_s$$

$$- \frac{1}{2} \int_0^t (M - N)_s \, d(M - N)_s + [M, N]_t.$$

Proof. The existence of the decomposition follows from (5.7), since the right member of that equation is a local martingale and $[M, N]$ has the properties stated above and its initial value is zero. Indeed, by substituting the expressions for $(M \pm N)^2 - [M \pm N]$ from (4.1) into (5.7), we obtain the above formula for the decomposition. The uniqueness follows from Corollary 4.5. ∎

If M and N are actually L^2-martingales, then by Theorem 4.6 the right side of (5.7) is a martingale, and $[M + N]$ and $[M - N]$ are integrable. In this case we have the following refinement of Theorem 5.2.

Theorem 5.3. *Let M and N be continuous L^2-martingales. Then there is a unique decomposition of MN as the sum of a continuous martingale and a continuous integrable process which is locally of bounded variation and has initial value zero. Moreover, for $0 \leq s < t$ we have*

$$(5.8) \quad \begin{aligned} E\left\{(M_t - M_s)(N_t - N_s) \mid \mathcal{F}_s\right\} &= E\left\{M_t N_t - M_s N_s \mid \mathcal{F}_s\right\} \\ &= E\left\{[M, N]_t - [M, N]_s \mid \mathcal{F}_s\right\}. \end{aligned}$$

Proof. The existence of the decomposition was explained above and the uniqueness follows from Theorem 5.2. The first equality in (5.8) follows on taking conditional expectations relative to $\mathcal{F}_s$ of the identity

$$(M_t - M_s)(N_t - N_s) = M_t N_t - M_s N_s - (M_t - M_s)N_s - (N_t - N_s)M_s.$$

The second equality in (5.8) follows from the fact that $MN - [M, N]$ is a martingale. ∎

Corollary 5.4. *Let M and N be continuous local martingales and τ be an optional time. For each t let $M_t^\tau = M_{t \wedge \tau}$ and $N_t^\tau = N_{t \wedge \tau}$. Then a.s. for each t we have*

$$(5.9) \quad [M^\tau, N^\tau]_t = [M, N]_{t \wedge \tau}.$$

In other words, this corollary states that the mutual variation process for two continuous local martingales stopped by the same optional time is indistinguishable from their mutual variation process stopped by that time.

Proof. By the decomposition of MN, $(M^\tau N^\tau)_t$ is the sum of a continuous local martingale evaluated at time $t \wedge \tau$, and $[M, N]_{t \wedge \tau}$. Then (5.9) follows by the uniqueness of the decomposition of $M^\tau N^\tau$. ∎

The following characterization of the mutual variation process plays a role in the proof of the multi-dimensional Itô formula.

Theorem 5.5. *Let M and N be continuous local martingales. For each t let $\{\pi_t^n, n \in I\!N\}$ be a sequence of partitions of $[0,t]$ such that $\lim_{n \to \infty} \delta\pi_t^n = 0$. Then, as $n \to \infty$,*

$$(5.10) \qquad \sum_j \left(M_{t_{j+1}} - M_{t_j}\right)\left(N_{t_{j+1}} - N_{t_j}\right) \to [M, N]_t \quad \text{in pr.,}$$

where the sum is over all j such that t_j and t_{j+1} are in π_t^n. If M and N are in fact L^2-martingales, then the convergence in (5.10) also holds in L^1. (The notation in (5.10) is the same as that used in Theorem 4.1).

Proof. From the definition of $[M, N]$ and Theorem 4.1 we have

$$[M, N]_t = \frac{1}{4}\left([M + N]_t - [M - N]_t\right)$$

$$= \lim_{n \to \infty} \frac{1}{4} \sum_j \left[\left\{\left(M_{t_{j+1}} + N_{t_{j+1}}\right) - \left(M_{t_j} + N_{t_j}\right)\right\}^2 \right.$$

$$\left. - \left\{\left(M_{t_{j+1}} - N_{t_{j+1}}\right) - \left(M_{t_j} - N_{t_j}\right)\right\}^2\right]$$

where the convergence is in pr. By Theorem 4.2(iii), the convergence also holds in L^1 if M and N are L^2-martingales. Simplifying the difference of squares in the above sum gives (5.10). ∎

The preceding result enables us to prove a kind of Cauchy-Schwarz inequality.

Notation. If M and N are continuous local martingales and $s < t$, we use $[M, N]_s^t$ to denote $[M, N]_t - [M, N]_s$; and $[M]_s^t$ and $[N]_s^t$ to denote $[M, M]_s^t$ and $[N, N]_s^t$, respectively. We use $\|[M, N]\|_s^t$ to denote the (total) variation of $[M, N]$ over the interval $[s, t]$. Note this should be distinguished from $|[M, N]_s^t|$, which denotes the magnitude of $[M, N]_s^t$ and is less than or equal to $\|[M, N]\|_s^t$.

Corollary 5.6. *Let M and N be continuous local martingales. Then for any $s < t$ in $I\!R_+$, we have almost surely,*

$$(5.11) \qquad \qquad |[M, N]|_s^t \leq \left([M]_s^t [N]_s^t\right)^{1/2}.$$

Proof. We first prove (5.11) with $|[M, N]_s^t|$ in place of $\|[M, N]\|_s^t$. For this, let $s < t$ in $\mathbb{R}_+$. Since convergence in pr. implies convergence of a subsequence a.s., it follows from Theorems 5.5 and 4.1(ii) that there is a sequence of partitions $\{\pi^n, n \in \mathbb{N}\}$ of $[s, t]$ (defined in the obvious way) with the mesh size of π^n tending to zero as $n \to \infty$, such that almost surely:

$$(5.12) \qquad [M, N]_s^t = \lim_{n \to \infty} \sum_j \left(M_{t_{j+1}} - M_{t_j}\right) \left(N_{t_{j+1}} - N_{t_j}\right)$$

and

$$\begin{cases} [M]_s^t = \lim_{n \to \infty} \sum_j \left(M_{t_{j+1}} - M_{t_j}\right)^2 \\ [N]_s^t = \lim_{n \to \infty} \sum_j \left(N_{t_{j+1}} - N_{t_j}\right)^2. \end{cases}$$

Applying Cauchy's inequality to the right member of (5.12) and then using the above, we obtain the following inequality almost surely,

$$(5.13) \qquad |[M, N]_s^t| \le \left([M]_s^t [N]_s^t\right)^{1/2}.$$

In fact, since the processes $[M], [N], [M, N]$ are all continuous, we may assume that this holds almost surely for all $s < t$, where the exceptional P-null set does not depend on s, t.

For the proof of (5.11), fix $s < t$ in $\mathbb{R}_+$ and let $\pi = \{t_j, j = 0, 1, \ldots, k\}$ be a partition of $[s, t]$. Then by (5.13), with t_j, t_{j+1} in place of s, t there, we have almost surely

$$\sum_{j=0}^{k-1} \left|[M, N]_{t_j}^{t_{j+1}}\right| \le \sum_{j=0}^{k-1} \left([M]_{t_j}^{t_{j+1}} [N]_{t_j}^{t_{j+1}}\right)^{1/2}.$$

By Cauchy's inequality and the additivity of $[M], [N]$ over disjoint intervals, it follows that almost surely,

$$\sum_{j=0}^{k-1} \left|[M, N]_{t_j}^{t_{j+1}}\right| \le \left(\sum_{j=0}^{k-1} [M]_{t_j}^{t_{j+1}}\right)^{1/2} \left(\sum_{j=0}^{k-1} [N]_{t_j}^{t_{j+1}}\right)^{1/2}$$
$$= \left([M]_s^t [N]_s^t\right)^{1/2}.$$

Since the exceptional P-null set on which the above may fail to hold does not depend on the points in the partition π, we obtain (5.11) by taking the supremum over all such partitions in the right member above. ∎

Next we obtain a formula for the mutual variation associated with two stochastic integrals.

Notation. If M is a continuous local martingale and $X \in \Lambda(\mathcal{P}, M)$, we use $X \cdot M$ to denote the process defined by

$$(X \cdot M)_t = \int_0^t X_s \, dM_s.$$

Theorem 5.7. *Let M and N be continuous local martingales, $X \in \Lambda(\mathcal{P}, M)$ and $Y \in \Lambda(\mathcal{P}, N)$. Then a.s. we have for all t:*

(5.14)
$$[X \cdot M, Y \cdot N]_t = \int_0^t X_s Y_s \, d[M, N]_s.$$

Proof. By replacing M and N by $M - M_0$ and $N - N_0$, respectively, and using a localizing sequence, we may suppose that M and N are continuous L^2-martingales and $X \in \Lambda^2(\mathcal{P}, M)$, and $Y \in \Lambda^2(\mathcal{P}, N)$.

Our first step is to verify that the right side of (5.14) is well-defined. Since X and Y are $\mathcal{B} \times \mathcal{F}$-measurable, then for each ω, $X_s(\omega)$ and $Y_s(\omega)$ are $\mathcal{B}$-measurable functions of s. Moreover, for almost every ω, the function $s \to [M, N]_s(\omega)$ is locally of bounded variation on $\mathbb{R}_+$. Then for such ω, the right member of (5.14) can be defined as a Lebesgue-Stieltjes integral whenever t is such that

(5.15)
$$\int_0^t |X_s(\omega)Y_s(\omega)| \, d\,|[M, N]|_s(\omega) < \infty,$$

where we have used $|[M, N]|_s$ as an abbreviation for the total variation process $|[M, N]|_0^s$. It follows from the next lemma that for almost every ω, (5.15) holds for all t in $\mathbb{R}_+$.

Lemma 5.8. *Let M and N be continuous L^2-martingales and suppose $X \in \Lambda^2(\mathcal{P}, M)$ and $Y \in \Lambda^2(\mathcal{P}, N)$. Then for each t we have*

(5.16)
$$E\left(\int_0^t |X_s Y_s|\, d\,|[M,N]|_s\right)$$
$$\leq \left\{\left(\int 1_{[0,t]}(X)^2\, d\mu_M\right)\left(\int 1_{[0,t]}(Y)^2\, d\mu_N\right)\right\}^{1/2}.$$

Proof. We first prove that for any measurable processes U and V we have a.s.

(5.17)
$$\int_0^t |U_s V_s|\, d\,|[M,N]|_s \leq \left\{\left(\int_0^t (U_s)^2\, d[M]_s\right)\left(\int_0^t (V_s)^2\, d[N]_s\right)\right\}^{1/2}.$$

In the above, some or all of the integrals may be infinite. The inequality (5.17) is a form of one due to Kunita and Watanabe.

By (5.11) we have a.s. for each $0 \leq u < s \leq t$:

(5.18)
$$|[M,N]|_u^s \leq ([M]_u^s [N]_u^s)^{\frac{1}{2}}.$$

Fix an ω at which (5.18) holds and the right member is finite for $0 \leq u < s \leq t$. By the measurability of U and V, $U_s(\omega)$ and $V_s(\omega)$ are Borel measurable functions of $s \in \mathbb{R}_+$. Consider first the case where

(5.19)
$$U_s(\omega) = \sum_{j=1}^n u_j 1_{[t_j, t_{j+1})}(s)$$
$$V_s(\omega) = \sum_{j=1}^n v_j 1_{[t_j, t_{j+1})}(s)$$

for $0 \leq t_0 < t_1 < \ldots < t_n \leq t$ and $u_j, v_j \in \mathbb{R}$ for $j = 1, \ldots, n$. Then by

(5.18) we have

$$\int_0^t |U_s V_s|\, d\,|[M,N]|_s\,(\omega) \le \sum_{j=1}^n |u_j v_j|\,|[M,N]|_{t_j}^{t_{j+1}}\,(\omega)$$

$$\le \sum_{j=1}^n |u_j v_j|\left([M]_{t_j}^{t_{j+1}}[N]_{t_j}^{t_{j+1}}\right)^{1/2}(\omega).$$

By Cauchy's inequality, the expression in the last line above is dominated by

$$\left\{\left(\sum_{j=1}^n u_j^2[M]_{t_j}^{t_{j+1}}\right)\left(\sum_{j=1}^n v_j^2[N]_{t_j}^{t_{j+1}}\right)\right\}^{1/2}(\omega),$$

which equals the right member of (5.17) evaluated at ω. Thus (5.17) holds at ω when $U(\omega)$ and $V(\omega)$ are given by (5.19). In particular, it holds when $U(\omega)$ and $V(\omega)$ are indicator functions of finite disjoint unions of intervals of the form $[u, v)$, $0 \le u < v \le t$. It then follows by a monotone class theorem (see Chung [11, Theorem 2.1.2]) that (5.17) holds whenever $U(\omega)$ and $V(\omega)$ are the indicator functions of Borel subsets of $[0, t)$. Thus an argument similar to that following (5.19) establishes (5.17) whenever $U(\omega)$ and $V(\omega)$ are Borel simple functions, i.e., when the indicators of intervals in (5.19) are replaced by indicators of disjoint Borel subsets of $[0, t)$. Finally, by a standard approximation argument in L^2-space, (5.17) holds at ω for all $U(\omega)$ and $V(\omega)$ for which the right member is finite. This extends to all $U(\omega)$ and $V(\omega)$ by monotone convergence. Since the set of admissible ω has probability one, it follows that (5.17) holds a.s. For an alternative method of verifying the extension of (5.17) from U and V of the form (5.19) to those for which the right member of (5.17) is finite, see Exercise 3 below.

By taking expectations in (5.17) with $U = X$ and $V = Y$, applying the Cauchy-Schwarz inequality, and then using Theorem 4.2(v) to write integrals with respect to quadratic variation processes as integrals with respect to corresponding Doléans measures, we obtain (5.16). ∎

We now return to the proof of Theorem 5.7. Since $X \in \Lambda^2(\mathcal{P}, M)$ and $Y \in \Lambda^2(\mathcal{P}, N)$, it follows from the above lemma that for each t the right side of (5.16) is finite and therefore (5.15) holds for almost every

ω. Moreover, since the integral in (5.15) is increasing in t, almost surely, (5.15) holds simultaneously for all t and then the right member of (5.14) is a well-defined continuous function of t. As a process it is locally of bounded variation and P-integrable for each t, by (5.16). Consequently, to prove (5.14) it suffices, by the uniqueness of the decomposition of $(X \cdot M)(Y \cdot N)$ given in Theorem 5.3, to show that

$$(5.20) \qquad \left\{ (X \cdot M)_t (Y \cdot N)_t - \int_0^t X_s Y_s \, d[M, N]_s, t \in \mathbb{R}_+ \right\}$$

is a martingale.

We first verify this for X and Y in $\mathcal{E}$, i.e., they are of the form (2.6). Since the process in (5.20) is separately linear in X and Y, it suffices to consider the cases in which X and Y are indicator functions of disjoint or identical predictable rectangles. If either X or Y is $1_{\{0\} \times F}$ for some $F \in \mathcal{F}_0$, it is easy to verify that the process in (5.20) is identically zero. Having dispensed with this case and invoking symmetry, it suffices to consider the following two cases:

(i) $X = 1_{(r,s] \times F}$ and $Y = 1_{(u,v] \times G}$ for some $0 \le r < s \le u < v < \infty, F \in \mathcal{F}_r$, and $G \in \mathcal{F}_u$;

(ii) $X = 1_{(r,s] \times F}$ and $Y = 1_{(r,s] \times G}$ for some $0 \le r < s < \infty$, and F and G in $\mathcal{F}_r$.

If (i) holds, then $\int_0^t X_z Y_z \, d[M, N]_z \equiv 0$ and

$$(X \cdot M)_t (Y \cdot N)_t = \begin{cases} 0 \text{ for } 0 \le t \le u \\ 1_{F \cap G}(M_s - M_r)(N_{t \wedge v} - N_u) \text{ for } t \ge u. \end{cases}$$

Moreover, $1_{F \cap G}(M_s - M_r) \in \mathcal{F}_u$ and $\{N_{t \wedge v} - N_u, t \ge u\}$ is a martingale with mean zero. It is then straightforward to verify that $(X \cdot M)(Y \cdot N)$ is a martingale and consequently, so too is (5.20).

If (ii) holds, then

$$(X \cdot M)_t (Y \cdot N)_t - \int_0^t X_z Y_z \, d[M, N]_z$$

$$= \begin{cases} 0 \text{ for } 0 \le t \le r \\ 1_{F \cap G} \{ (M_{t \wedge s} - M_r)(N_{t \wedge s} - N_r) \\ \quad -([M, N]_{t \wedge s} - [M, N]_r) \} \text{ for } t \ge r \end{cases}$$

where the term in braces can be expressed as the sum of the three martingales $\{M_{t \wedge s} N_{t \wedge s} - [M, N]_{t \wedge s}, t \ge r\}$, $\{-N_r M_{t \wedge s}, t \ge r\}$, and $\{-M_r N_{t \wedge s}, t \ge r\}$, plus the term $M_r N_r + [M, N]_r$. With this, case (ii) is settled.

Hence (5.20) is a martingale whenever X and Y are in $\mathcal{E}$.

More generally, for $X \in \Lambda^2(\mathcal{P}, M)$ and $Y \in \Lambda^2(\mathcal{P}, N)$, for each fixed $T \in \mathbb{R}_+$, there is a sequence $\{X^n\}$ in $\mathcal{E}$ such that $1_{[0,T]} X^n \to 1_{[0,T]} X$ in $\mathcal{L}^2(\mu_M)$ as $n \to \infty$ and the same is true with Y and N in place of X and M, respectively. By what we have just proved,

$$\left\{ (X^n \cdot M)_t (Y^n \cdot N)_t - \int_0^t X_s^n Y_s^n \, d[M, N]_s, t \in [0, T] \right\}$$

is a martingale for each n. By using the $\mathcal{L}^2$ convergence cited above, the isometry (2.10), and (5.16), it can be verified that for each $t \in [0, T]$, the above expression converges in L^1 to that in (5.20). Since the martingale property is preserved by L^1-limits, it follows that (5.20) is a martingale.

∎

Corollary 5.9. *Let M be a continuous local martingale and suppose $X \in \Lambda(\mathcal{P}, M)$. Then a.s. we have for all t:*

(5.21) $$[X \cdot M]_t = \int_0^t (X_s)^2 \, d[M]_s.$$

In the following example, we determine the mutual variation process for a d-dimensional Brownian motion.

Example. Let $\mathbf{B} = (B^1, \ldots, B^d)$ be a Brownian motion in $\mathbb{R}^d$. We have already seen in the discussion of quadratic variation that $[B^i, B^i]_t = t$ for $i = 1, \ldots, d$. We shall show here that $[B^i, B^j] = 0$ for $i \neq j$. Since the mutual variation is the same for $\mathbf{B} - \mathbf{B}_0$ as $\mathbf{B}$, we may assume without loss of generality that $\mathbf{B}_0 = 0$ a.s. Then for $i \neq j$ and $s < t$ in $\mathbb{R}_+$,

$$
\begin{aligned}
E &\left\{ B_t^i B_t^j - B_s^i B_s^j \mid \mathcal{F}_s \right\} \\
&= E \left\{ (B_t^i - B_s^i)(B_t^j - B_s^j) \mid \mathcal{F}_s \right\} \\
&\quad + E \left\{ B_s^i (B_t^j - B_s^j) \mid \mathcal{F}_s \right\} + E \left\{ B_s^j (B_t^i - B_s^i) \mid \mathcal{F}_s \right\} \\
&= E \left\{ (B_t^i - B_s^i)(B_t^j - B_s^j) \right\} \\
&\quad + B_s^i E \left\{ B_t^j - B_s^j \mid \mathcal{F}_s \right\} + B_s^j E \left\{ B_t^i - B_s^i \mid \mathcal{F}_s \right\},
\end{aligned}
$$

where we have used the independence of the increments of $\mathbf{B}$. The first term in the last two lines of the above equation is zero because the components of $\mathbf{B}$ are independent one-dimensional Brownian motions whose increments have zero mean. The second term is zero because $B_t^j - B_s^j$ has zero mean and is independent of $\mathcal{F}_s$, and similarly the third term is zero. It follows that for $i \neq j$, $B^i B^j$ is a martingale and hence by the decomposition theorem 5.3, $[B^i, B^j] = 0$. ∎

Remark. For a generalization of the result that $B^i B^j$ is a continuous local martingale for $i \neq j$, see Exercise 6.

5.4 Multi-Dimensional Itô Formula

Theorem 5.10. *Let $m \in \mathbb{N}$ and $n \in \mathbb{N}$. Let M^i be a continuous local martingale for $1 \leq i \leq m$, and V^k be a continuous process which is locally of bounded variation for $1 \leq k \leq n$. Suppose that D is a domain in $\mathbb{R}^{m+n}$ such that a.s.*

$$
Z_t \equiv \left(M_t^1, \ldots, M_t^m, V_t^1, \ldots, V_t^n \right)
$$

takes values in D for all t. Let $f(x, y)$ be a continuous real-valued function of $(x, y) \in D$ such that $\frac{\partial f}{\partial x_i}$, $\frac{\partial^2 f}{\partial x_i \partial x_j}$, $1 \leq i, j \leq m$, and $\frac{\partial f}{\partial y_k}$, $1 \leq k \leq n$, exist and are continuous in D. Then a.s. we have for all t:

$$f(Z_t) - f(Z_0) = \sum_{i=1}^{m} \int_0^t \frac{\partial f}{\partial x_i}(Z_s)\, dM_s^i$$

(5.22)
$$+ \sum_{k=1}^{n} \int_0^t \frac{\partial f}{\partial y_k}(Z_s)\, dV_s^k$$

$$+ \frac{1}{2} \sum_{i=1}^{m} \sum_{j=1}^{m} \int_0^t \frac{\partial^2 f}{\partial x_i \partial x_j}(Z_s)\, d[M^i, M^j]_s.$$

Sketch of proof. The method of proof is similar to that for the one-dimensional formula. The main differences are that a theorem analogous to Theorem 4.8 must be proved for mutual variation using (5.10); and the functions g_n with bounded partial derivatives which are used to approximate f must be chosen to have supports contained in D_n where $\{D_n\}$ is a sequence of subdomains of D such that $D_n \subset D_{n+1}$, $\overline{D}_n$ is a compact subset of D, and $D = \bigcup_n D_n$. ∎

Remark. By the continuity of $t \to f(Z_t)$, the constant time t in (5.22) can be replaced by a random time τ. For example,

$$\tau_E = \inf\{s \geq 0 : Z_s \notin E\},$$

the first exit time from a subdomain E of D with $\overline{E} \subset D$ is such a time. This is often used in applications.

It is worthwhile stating the Itô formula for Brownian motion as a corollary.

Corollary 5.11. *Let* $\mathbf{B} = (B^1, \ldots, B^d)$ *be a Brownian motion in* $\mathbb{R}^d$. *Suppose that D is a domain in $\mathbb{R}^d$ such that*

$$P(\mathbf{B}_t \in D \text{ for all } t) = 1.$$

Let f be a continuous real-valued function on D such that $\frac{\partial f}{\partial x_i}$ *and* $\frac{\partial^2 f}{\partial x_i \partial x_j}$

exist and are continuous in D for $1 \leq i, j \leq d$. Then a.s. we have for all t:

$$(5.23) \qquad f(\mathbf{B}_t) - f(\mathbf{B}_0) = \sum_{i=1}^{d} \int_0^t \frac{\partial f}{\partial x_i}(\mathbf{B}_s)\, dB_s^i + \frac{1}{2} \int_0^t \Delta f(\mathbf{B}_s)\, ds$$

where Δf denotes the Laplacian of f, defined by $\Delta f = \sum_{i=1}^{d} \frac{\partial^2 f}{\partial x_i^2}$.

Example. If Corollary 5.11 is applied to a Brownian motion in $I\!R^3$ starting at $x_0 \neq 0$, with $D = I\!R^3 \backslash \{0\}$ and $f(x) = |x|^{-1}$ where $x = (x_1, x_2, x_3)$; then since $P^{x_0}(\mathbf{B}_t = 0$ for any $t \geq 0) = 0$, $\frac{\partial f}{\partial x_i} = -\frac{x_i}{|x|^3}$ and $\Delta f = 0$ in D, (5.23) becomes

$$\frac{1}{|\mathbf{B}_t|} - \frac{1}{|\mathbf{B}_0|} = -\sum_{i=1}^{3} \int_0^t \frac{B_s^i}{|\mathbf{B}_s|^3}\, dB_s^i.$$

Each of the stochastic integrals in the sum on the right is a local martingale and hence $\left\{ |\mathbf{B}_t|^{-1}, t \in I\!R_+ \right\}$ is a local martingale; a result which was obtained at the end of Chapter 1 by more direct means.

Example. *(Bessel Process.)* Let $\mathbf{B}$ be a Brownian motion in $I\!R^d$, $d \geq 2$, such that $\mathbf{B}(0) = x_0 \in I\!R^d \backslash \{0\}$ *P*-a.s. Then $P(\mathbf{B}_t = 0$ for some $t \geq 0) = 0$ (cf. Exercise 12 of this chapter). By applying Corollary 5.11 with $D = I\!R^d \backslash \{0\}$ and $f(x) = |x|$, so that $\frac{\partial f}{\partial x_i} = x_i / |x|$ and $\Delta f = (d-1)/|x|$, we obtain a.s. for all $t \geq 0$:

$$|\mathbf{B}_t| - |x_0| = \int_0^t |\mathbf{B}_s|^{-1} \mathbf{B}_s\, d\mathbf{B}_s + \int_0^t \left(\frac{d-1}{2} \right) |\mathbf{B}_s|^{-1}\, ds,$$

where $\int_0^t |\mathbf{B}_s|^{-1} \mathbf{B}_s \cdot d\mathbf{B}_s \equiv \sum_{i=1}^{d} \int_0^t |\mathbf{B}_s|^{-1} B_s^i dB_s^i$ defines a continuous local martingale $\{W_t, t \geq 0\}$. By (5.14) and the fact that $[B^i, B^j]_t = \delta_{ij} t$, we have

$$[W]_t = \sum_{i=1}^{d} \int_0^t |\mathbf{B}_s|^{-2} (B_s^i)^2 ds = t.$$

It follows from the characterization of Brownian motion given in Theorem 6.1 of the next chapter that W is a one-dimensional Brownian motion.

Thus, $R_t \equiv |\mathbf{B}_t|$ is a (weak) solution of the *stochastic differential equation*:

$$dR_t = \left(\frac{d-1}{2}\right) R_t^{-1} dt + dW_t,$$

where W is a one-dimensional Brownian motion.

Remark. Any solution of this stochastic differential equation satisfying $R_0 = r_0 \in (0, \infty)$ is uniquely determined in law (cf. Chapter 10) and is said to be a representation of the *Bessel process* that starts from r_0 and has parameter $\frac{d}{2} - 1$. The parameter is $\frac{d}{2} - 1$ because this is the order of the modified Bessel function appearing in the transition density for the Bessel process (see for example Karlin and Taylor [50, pp. 367–371]).

5.5 Exercises

1. In this exercise, B is a Brownian motion on $\mathbb{R}$ with $B_0 = 0$ a.s.
(a) Let f be a function of t alone. Suppose f has compact support in $\mathbb{R}_+$ and is of bounded variation. Prove that:

$$\int_0^\infty f(t) dB(t) = -\int_0^\infty B(t) df(t) \quad \text{a.s.}$$

(b) Suppose f_1 and f_2 are continuous functions of t alone. Prove that for each $t \geq 0$:

$$\int_0^t f_2(s) \left[\int_0^s f_1(u) dB(u)\right] ds = \int_0^t \left[\int_u^t f_2(s) ds\right] f_1(u) dB(u) \quad \text{a.s.}$$

2. Show that if M is a continuous local martingale and V is a continuous process that is locally of bounded variation, then a.s. for all $t \in \mathbb{R}_+$:

$$M_t V_t = M_0 V_0 + \int_0^t V_s dM_s + \int_0^t M_s dV_s.$$

This is the same as the integration by parts formula for ordinary calculus.

3. Use Lusin's theorem (see Royden [69]) in place of a monotone class argument to extend (5.17) from the case where $U(\omega)$ and $V(\omega)$ satisfy (5.19) to that where $U(\omega)$ and $V(\omega)$ are square integrable with respect to $d[M](\omega)$ and $d[N](\omega)$, respectively.

Hint: First extend (5.17) to the case in which $U(\omega)$ and $V(\omega)$ are continuous on $[0, t]$ and then apply Lusin's theorem with the measure $\mu \equiv \mu_0 + \mu_1 + \mu_2$ on $[0, t]$, where μ_0 is the measure induced there by the total variation of $[M, N](\omega)$, μ_1 is the measure induced by $d[M](\omega)$ and μ_2 is that induced by $d[N](\omega)$.

4. The fact that (5.17) follows from (5.18) is a consequence of the following *generalized Cauchy-Schwarz inequality*. Prove this inequality.

Suppose μ, μ_1 and μ_2 are finite (non-negative) measures on $[0, t)$ satisfying for $I = [a, b) \subset [0, t)$,

$$\mu(I)^2 \leq \mu_1(I)\mu_2(I).$$

Then for positive Borel measurable functions f and g,

$$\left(\int f g d\mu \right)^2 \leq \int f^2 d\mu_1 \int g^2 d\mu_2.$$

5. Let M and N be continuous local martingales and let Y be a bounded continuous adapted process. Let $t \in \mathbb{R}_+$ and $\{\pi_t^n, n \in \mathbb{N}\}$ be a sequence of partitions of $[0, t]$ such that $\lim_{n \to \infty} \delta \pi_t^n = 0$. For each $n \in \mathbb{N}$, let

$$Z_n = \sum_j Y_{t_j}(M_{t_{j+1}} - M_{t_j})(N_{t_{j+1}} - N_{t_j})$$

where the sum is over all j such that t_j and t_{j+1} are both in π_t^n. Prove that $\{Z_n, n \in \mathbb{N}\}$ converges in probability to $\int_0^t Y_s d[M, N]_s$.

6. Suppose M and N are two *independent* integrable processes. For each $s \geq 0$, let $\mathcal{G}_s^1 = \sigma\{M_u : 0 \leq u \leq s\}$, $\mathcal{G}_s^2 = \sigma\{N_u : 0 \leq u \leq s\}$, and $\mathcal{G}_s \equiv \sigma\{M_u, N_u : 0 \leq u \leq s\}$. Then for any times s, t_1, and t_2, we have

$$E\{M_{t_1} N_{t_2} \mid \mathcal{G}_s\} = E\{M_{t_1} \mid \mathcal{G}_s^1\} E\{N_{t_2} \mid \mathcal{G}_s^2\}.$$

Deduce from this that if M and N are continuous martingales, then MN is a continuous martingale and $[M, N] \equiv 0$.

Hint: First verify that the left member of the above equality is well defined. Then use the fact that the σ-field $\mathcal{G}_s^1$ is generated by the collection of sets of the form $\{M_{u_1} \in A_1, \ldots, M_{u_n} \in A_n\}$ where $0 \leq u_1 < \ldots < u_n \leq s$, $A_1, \ldots, A_n$ are Borel subsets of $\mathbb{R}$, and $n \in \mathbb{N}$; and similarly for $\mathcal{G}_s^2$ and $\mathcal{G}_s$.

7. Suppose that M^1 and M^2 are continuous local martingales and V^1 and V^2 are continuous processes that are locally of bounded variation. Define *semimartingales* $X^i = M^i + V^i$ for $i = 1, 2$. Let $t \in \mathbb{R}_+$ and $\{\pi_t^n, n \in \mathbb{N}\}$ be a sequence of partitions of $[0, t]$ as in Theorem 5.5. Prove that as $n \to \infty$, for $i, j \in \{1, 2\}$,

$$\sum_{t_k, t_{k+1} \in \pi_t^n} (X_{t_{k+1}}^i - X_{t_k}^i)(X_{t_{k+1}}^j - X_{t_k}^j) \to < M^i, M^j >_t \quad \text{in pr.}$$

8. Fix $-\infty < a < b < \infty$ and let B be a one-dimensional Brownian motion starting from $x \in (a, b)$. Let P^x and E^x denote the associated probability and expectation. Let $\tau = \inf\{t \geq 0 : B_t \notin (a, b)\}$. Recall that $B_t^2 - t$ defines a continuous martingale. Combine this with Doob's stopping theorem to prove that:

$$E^x[\tau] < \infty.$$

It follows from this that

$$P^x(\tau < \infty) = 1.$$

In the following five exercises, **B** denotes a d-dimensional Brownian motion, and P^x and E^x denote the probability and expectation, respectively, associated with **B** starting from $x \in \mathbb{R}^d$.

9. Use the result from Exercise 8 to show that

$$P^x(\tau_R < \infty) = 1,$$

where $\tau_R = \inf\{t \geq 0 : |\mathbf{B}_t| \geq R\}$ for $R \in \mathbb{R}_+$.

10. Let D be a *bounded* domain in $\mathbb{R}^d$ and suppose f is a real-valued function that is twice-continuously differentiable in some domain containing

$\overline{D}$. Let $\tau = \inf\{t \geq 0 : \mathbf{B}_t \notin D\}$. Use the multi-dimensional Itô formula to prove that for each $x \in \overline{D}$, under P^x,

$$f(\mathbf{B}_{t \wedge \tau}) - \frac{1}{2} \int\limits_0^{t \wedge \tau} \Delta f(\mathbf{B}_s) ds, \quad t \geq 0,$$

defines a martingale. If in addition, f is harmonic in D, show that f has the following representation:

$$f(x) = E^x[f(B_\tau)] \quad \text{for all } x \in \overline{D}.$$

11. Use the results of Exercises 9 and 10, together with the function f defined by

$$f(x) = \sum_{i=1}^d x_i^2, \quad x \in \mathbb{R}^d,$$

to obtain the following

$$E^x[\tau_R] = \frac{R^2 - |x|^2}{d},$$

for $0 \leq |x| \leq R < \infty$, where τ_R is defined as in Exercise 9.

12. Apply the result of Exercise 10 to the function f defined by

$$f(x) = \begin{cases} x & \text{for } d = 1 \\ \ln|x| & \text{for } d = 2 \\ |x|^{2-d} & \text{for } d \geq 3, \end{cases}$$

and then use the result of Exercise 9 to conclude the following.

For $d = 1$, $-\infty < a \leq x \leq b < \infty$, $\tau_a = \inf\{t \geq 0 : B_t \leq a\}$ and $\tau_b = \inf\{t \geq 0 : B_t \geq b\}$:

$$P^x(\tau_a < \tau_b) = \frac{b - x}{b - a},$$

and $P^x(\tau_a < \infty) = 1$, $P^x(\tau_b < \infty) = 1$.

For $d \geq 2$, let $0 \leq r \leq |x| \leq R < \infty$ and define $\tau_r = \inf\{t \geq 0 : |\mathbf{B}_t| \leq r\}$ and $\tau_R = \inf\{t \geq 0 : |\mathbf{B}_t| \geq R\}$. Then for $r > 0$,

$$P^x(\tau_r < \tau_R) = \begin{cases} (\ln R - \ln |x|)/(\ln R - \ln r) & \text{if } d = 2 \\ (|x|^{2-d} - R^{2-d})/(r^{2-d} - R^{2-d}) & \text{if } d \geq 3. \end{cases}$$

Letting $r \to 0$ in the above yields

$$P^x(\tau_0 < \tau_R) = 0 \quad \text{for} \quad 0 < |x| \leq R < \infty,$$

and then letting $R \to \infty$, from the continuity of the sample paths of Brownian motion, we obtain

$$P^x(\tau_0 < \infty) = 0 \quad \text{for all} \quad x \in \mathbb{R}^d \backslash \{0\}.$$

Now when $d \geq 3$, letting $R \to \infty$ in the above expression for $P^x(\tau_r < \tau_R)$ yields

$$P^x(\tau_r < \infty) = \left(\frac{|x|}{r}\right)^{2-d} \quad \text{for} \quad 0 < r \leq |x| < \infty.$$

Combine this with the result of Exercise 9 and the strong Markov property of Brownian motion to conclude that Brownian motion is transient in $\mathbb{R}^d$ for $d \geq 3$, i.e., for each $x \in \mathbb{R}^d$, $|\mathbf{B}_t| \to \infty$ P^x-a.s. as $t \to \infty$.

13. Use Corollary 5.11 to show that if u is harmonic (i.e., $\Delta u = 0$) and bounded on $\mathbb{R}^d$, then $\{u(\mathbf{B}_t), t \geq 0\}$ defines a P^x-martingale for each $x \in \mathbb{R}^d$. Then use the martingale convergence theorem (Theorem 1.5) and the triviality of the tail σ-field of Brownian motion (i.e., if $A \in \sigma\{\mathbf{B}_s : s \geq t\}$ for all $t \geq 0$, then $P^x(A) = 0$ for all x or $= 1$ for all x), to prove Liouville's theorem: any bounded harmonic function in $\mathbb{R}^d$ is constant.

6

APPLICATIONS OF THE
ITO FORMULA

6.1 Characterization of Brownian Motion

Theorem 6.1. *A process $M = \{M_t, t \in \mathbb{R}_+\}$ is a Brownian motion in $\mathbb{R}$ if and only if there is a standard filtration $\{\mathcal{F}_t\}$ such that $\{M_t, \mathcal{F}_t, t \in \mathbb{R}_+\}$ is a continuous local martingale with quadratic variation $[M]$ satisfying*

(6.1) $[M]_t = t$ *a.s. for all t.*

Proof. For the "only if" part, let M be a Brownian motion in $\mathbb{R}$ and let $\{\mathcal{F}_t\}$ be the usual standard filtration associated with this Brownian motion (see Section 1.8). Then $\{M_t, \mathcal{F}_t, t \in \mathbb{R}_+\}$ is a continuous local martingale, and by the example following Theorem 4.1, its quadratic variation is given by (6.1).

For the "if" part, suppose $\{M_t, \mathcal{F}_t, t \in \mathbb{R}_+\}$ is a continuous local martingale such that (6.1) holds. In the remainder of this proof, the ambient filtration $\{\mathcal{F}_t\}$ will be fixed and explicit mention of it will be omitted. Let $\{\tau_k\}$ be a localizing sequence for M such that $\{M_{t \wedge \tau_k} - M_0, t \in \mathbb{R}_+\}$ is an L^2-martingale for each k. Then by applying (4.1) to $M - M_0$, since

$[M - M_0] = [M]$, we obtain

$$(M_{t \wedge \tau_k} - M_0)^2 = 2 \int_0^{t \wedge \tau_k} (M_s - M_0)\, d(M_s - M_0) + [M]_{t \wedge \tau_k}.$$

By taking expectations, using Theorem 4.2(ii) and the assumption on $[M]$, we conclude that

$$E\left\{(M_{t \wedge \tau_k} - M_0)^2\right\} = E(t \wedge \tau_k) \leq t.$$

Hence the sup over k of the left member in the above is bounded and consequently $\{M_{t \wedge \tau_k} - M_0, k \in I\!N\}$ is uniformly integrable, by Proposition 1.1. It follows by Proposition 1.8 that $M - M_0$ is a martingale and by applying Fatou's lemma to the above that $M_t - M_0 \in L^2$ for each t. Thus $M - M_0$ is an L^2-martingale.

For each $\alpha \in I\!R$, let ϕ_α be the complex-valued function defined on $I\!R^2$ by $\phi_\alpha(x, y) = \exp\left(i\alpha x + \frac{1}{2}\alpha^2 y\right)$ for all (x, y) in $I\!R^2$. By applying Theorem 5.1 to the real and imaginary parts of ϕ_α separately, we see that the Itô formula yields a.s.

$$\phi_\alpha(M_t - M_0, t) - \phi_\alpha(0, 0) = \int_0^t \frac{\partial \phi_\alpha}{\partial x}(M_u - M_0, u)\, d(M_u - M_0)$$

$$+ \int_0^t \frac{\partial \phi_\alpha}{\partial y}(M_u - M_0, u)\, du$$

$$+ \frac{1}{2} \int_0^t \frac{\partial^2 \phi_\alpha}{\partial x^2}(M_u - M_0, u)\, d[M]_u.$$

Then, since $d[M]_u = du$ and $\frac{\partial \phi_\alpha}{\partial y} = -\frac{1}{2}\frac{\partial^2 \phi_\alpha}{\partial x^2}$, the above reduces to

(6.2)
$$\exp\left(i\alpha(M_t - M_0) + \frac{1}{2}\alpha^2 t\right) - 1$$

$$= i\alpha \int_0^t \exp\left(i\alpha(M_u - M_0) + \frac{1}{2}\alpha^2 u\right) d(M_u - M_0).$$

The above will be used to show for r and $s \geq 0$ that $M_{s+r} - M_s$ is independent of $\mathcal{F}_s$ and is a normally distributed r.v. with mean zero and variance r.

In the following, a complex-valued function N defined on $\mathbb{R}_+ \times \Omega$ will be called a martingale if the real and imaginary parts of N are martingales. In a similar fashion, the (conditional) expectation of a complex-valued function on Ω will be the sum of the (conditional) expectation of its real part and i times that of its imaginary part, whenever both are defined.

Since $M - M_0$ is an L^2-martingale and $\exp\left(i\alpha(M - M_0) + \frac{1}{2}\alpha^2 t\right)$ is bounded on compact intervals of t, it follows from Theorem 2.5 that the stochastic integral in (6.2) defines a martingale and hence $\{\exp\left(i\alpha(M_t - M_0) + \frac{1}{2}\alpha^2 t\right), t \geq 0\}$ is a martingale. Thus for r and $s \geq 0$,

(6.3)
$$E\left[\exp\left(i\alpha(M_{s+r} - M_0) + \frac{1}{2}\alpha^2(s + r)\right) \Big| \mathcal{F}_s\right]$$
$$= \exp\left(i\alpha(M_s - M_0) + \frac{1}{2}\alpha^2 s\right)$$

and therefore

(6.4)
$$E\left[\exp\left(i\alpha(M_{s+r} - M_s)\right) | \mathcal{F}_s\right] = \exp\left(-\frac{1}{2}\alpha^2 r\right)$$

Let $Y = M_{s+r} - M_s$, $Z \in \mathcal{F}_s$ and $\beta \in \mathbb{R}$. Then by (6.4),

(6.5)
$$E\left(\exp(i\alpha Y)\right) = \exp\left(-\frac{1}{2}\alpha^2 r\right)$$

and

$$E\left[\exp\left(i(\alpha Y + \beta Z)\right)\right] = \exp\left(-\frac{1}{2}\alpha^2 r\right) E\left(\exp(i\beta Z)\right)$$
$$= E\left(\exp(i\alpha Y)\right) E\left(\exp(i\beta Z)\right).$$

It follows from the above (see Chung [11, p. 187]) that Y and Z are independent, and hence Y is independent of $\mathcal{F}_s$.

Thus we have shown for each r and $s \geq 0$ that $M_{s+r} - M_s$ is independent of $\mathcal{F}_s$ and its characteristic function is $\alpha \rightarrow \exp\left(-\frac{1}{2}\alpha^2 r\right)$. Hence it is a normal r.v. with mean zero and variance r. It follows that M has the properties (i) and (ii) defining Brownian motion in $\mathbb{R}$ (see Section 1.8). ∎

6.2 Exponential Processes

In the next application of the Itô formula, we prove that for a continuous local martingale M with quadratic variation process A, the exponential process $Z^\alpha = \{\exp(\alpha M_t - \frac{1}{2}\alpha^2 A_t), t \in I\!R_+\}$ is a continuous local martingale for each $\alpha \in I\!R$. The converse of this result is also proved, although the proof does not employ the Itô formula. Furthermore, we give conditions under which the qualifier "local" can be omitted.

Theorem 6.2. *Let M and A be continuous adapted processes such that A is increasing and $A_0 = 0$. For each $\alpha \in I\!R$, let Z^α be the process defined by*

$$Z_t^\alpha = \exp\left(\alpha M_t - \frac{1}{2}\alpha^2 A_t\right).$$

Then the following two assertions are equivalent.

(i)   *M is a local martingale and $[M] = A$.*

(ii)   *For each $\alpha \in I\!R$, Z^α is a local martingale.*

Moreover, if M is an L^2-martingale with $[M] = A$ and α is such that $Z_0^\alpha \in L^2$ and

(6.6)
$$E\left(\int_0^t (Z_s^\alpha)^2\, dA_s\right) < \infty \text{ for each } t,$$

then Z^α is an L^2-martingale. On the other hand, if the following two conditions are satisfied:

(a)   *the random variable A_t is bounded for each t,*

(b)   *there is $\alpha_0 > 0$ such that $E(\exp(\alpha_0 |M_t|)) < \infty$ for each t and Z^α is a martingale for $|\alpha| \le \frac{1}{2}\alpha_0$,*

then M is an L^2-martingale with $[M] = A$.

Proof. Suppose (i) holds. To prove that (ii) follows, we apply the Itô formula to $f(x,y) = \exp\left(\alpha x - \frac{1}{2}\alpha^2 y\right)$, to obtain a.s.:

$$f(M_t, A_t) - f(M_0, A_0) = \int\limits_0^t \alpha f(M_s, A_s)\, dM_s + \int\limits_0^t \left\{ -\frac{\alpha^2}{2} f(M_s, A_s) \right\} dA_s$$

$$+ \frac{1}{2} \int\limits_0^t \alpha^2 f(M_s, A_s)\, dA_s.$$

This simplifies to

(6.7)
$$Z_t^\alpha - Z_0^\alpha = \alpha \int\limits_0^t Z_s^\alpha\, dM_s.$$

Since Z^α is a continuous adapted process, it follows from Theorem 2.11 that the right side of (6.7) is a local martingale and hence so is Z^α. If M is actually an L^2-martingale and (6.6) holds, then it follows from (4.5) that $Z^\alpha \in \Lambda^2(\mathcal{P}, M)$ and consequently the stochastic integral in (6.7) is an L^2-martingale; hence so is Z^α when $Z_0^\alpha \in L^2$. A condition which implies (6.6) is the following:

$$E\left(\int\limits_0^t \exp\left(2\alpha M_s \right) dA_s \right) < \infty \quad \text{for each } t.$$

For the proof that (ii) implies (i), we employ the following lemma which gives sufficient conditions for M to be an L^2-martingale.

Lemma 6.3. *Suppose conditions (a) and (b) hold. Then M is an L^2-martingale with $[M] = A$.*

Proof. For $0 \le s < t$, $F \in \mathcal{F}_s$, and $|\alpha| < \frac{1}{2}\alpha_0$, since Z^α is a martingale we have

(6.8)
$$\int\limits_F \exp\left(\alpha M_s - \frac{1}{2}\alpha^2 A_s \right) dP = \int\limits_F \exp\left(\alpha M_t - \frac{1}{2}\alpha^2 A_t \right) dP.$$

Now, there is a constant $K > 0$ such that for all $x \in I\!R$, $y \in I\!R_+$, $|\alpha| < \alpha_0/2$ and $n = 1, 2$,

$$|x|^n \le K \exp\left(\alpha_0 |x| / 2\right),$$

(6.9)
$$\exp\left(\alpha x - \frac{1}{2}\alpha^2 y\right) \le \exp\left(\alpha_0 |x| / 2\right).$$

Then by these inequalities, the positivity and boundedness of A_s and A_t, and the fact that $E\left[\exp\left(\alpha_0 |M_r|\right)\right]$ is finite for $r = s$ and $r = t$, we are justified in differentiating twice with respect to α under the integral signs in (6.8). Differentiating once gives

$$\int_F (M_s - \alpha A_s) \exp\left(\alpha M_s - \frac{1}{2}\alpha^2 A_s\right) dP$$
$$= \int_F (M_t - \alpha A_t) \exp\left(\alpha M_t - \frac{1}{2}\alpha^2 A_t\right) dP$$

and twice gives

$$\int_F \{(M_s - \alpha A_s)^2 - A_s\} \exp\left(\alpha M_s - \frac{1}{2}\alpha^2 A_s\right) dP$$
$$= \int_F \{(M_t - \alpha A_t)^2 - A_t\} \exp\left(\alpha M_t - \frac{1}{2}\alpha^2 A_t\right) dP.$$

By setting $\alpha = 0$ in the above expressions we obtain

(6.10)
$$\int_F M_s \, dP = \int_F M_t \, dP$$

and

(6.11)
$$\int_F \{(M_s)^2 - A_s\} \, dP = \int_F \{(M_t)^2 - A_t\} \, dP.$$

It follows from (6.10), (6.9) and condition (b), that M is an L^2-martingale, and from (6.11) that $(M)^2 - A$ is a martingale. Hence by the uniqueness of the decomposition of $(M)^2$ given in Theorem 4.6 we conclude that $[M] = A$.

∎

We now return to the proof that (ii) implies (i). Suppose Z^α is a local martingale for each $\alpha \in I\!R$. For each $k \in I\!N$, let

$$\tau_k = \inf \{t \geq 0 : |M_t| \vee |A_t| > k\},$$

$$M_t^k = M_{t \wedge \tau_k} - M_0$$

and

$$A_t^k = A_{t \wedge \tau_k}.$$

Then $Z_{t \wedge \tau_k}^\alpha - Z_0^\alpha$ is a bounded martingale and consequently so is

$$Z_t^{\alpha,k} \equiv \exp\left(\alpha M_t^k - \frac{1}{2}\alpha^2 A_t^k\right)$$
$$= (Z_{t \wedge \tau_k}^\alpha - Z_0^\alpha) \exp(-\alpha M_0) 1_{\{\tau_k > 0\}} + 1.$$

Since M^k and A^k are bounded, by applying Lemma 6.3 to $Z^{\alpha,k}$, we may conclude that M^k is an L^2-martingale with $[M^k] = A^k$. It follows that M is a local martingale with $\{\tau_k\}$ as a localizing sequence. Moreover, since $M_t^k = (M - M_0)_{t \wedge \tau_k}$ and $[M - M_0] = [M]$, it follows from Corollary 5.4 that $[M^k]_t = [M]_{t \wedge \tau_k}$. Letting $k \to \infty$ and using $[M^k] = A^k$, we obtain $A = [M]$. ∎

6.3 A Family of Martingales Generated by M

In the proof of Lemma 6.3, we saw under the conditions stated there that

$$\left.\frac{dZ^\alpha}{d\alpha}\right|_{\alpha=0} = M \quad \text{and} \quad \left.\frac{d^2 Z^\alpha}{d\alpha^2}\right|_{\alpha=0} = M^2 - A$$

are martingales. We next extend these results to higher derivatives. This provides us with a mechanism for generating polynomials in M and A which are martingales.

Notation. For each $n \in I\!N_0$, let $H_n(x,y)$ denote the polynomial function of x and y defined by

$$H_n(x,y) = \left.\frac{d^n}{d\alpha^n} \exp\left(\alpha x - \frac{1}{2}\alpha^2 y\right)\right|_{\alpha=0}.$$

Then

$$\exp\left(\alpha x - \frac{1}{2}\alpha^2 y\right) = \sum_{n=0}^{\infty} \frac{\alpha^n}{n!} H_n(x,y)$$

for all α in $I\!\!R$.

Theorem 6.4. *Let M and A be continuous adapted processes such that A is increasing and $A_0 = 0$. Suppose conditions (a) and (b) of Theorem 6.2 are satisfied. Then for each $n \in I\!\!N_0$, $H_n(M, A)$ is an L^2-martingale.*

Proof. For $n = 0$, $H_0(M, A) \equiv 1$ is clearly an L^2-martingale. Let $n \in I\!\!N$. Then there is a constant $K_n > 0$ such that for all $x \in I\!\!R$,

$$|x|^m \le K_n \exp\left(\alpha_0 |x|/2\right) \quad \text{for} \quad m = 1, 2, \ldots, n.$$

It then follows in a similar manner to that in the proof of Lemma 6.3, by differentiating under the integral signs in (6.8) and setting $\alpha = 0$, that

$$\left. \frac{d^n Z^\alpha}{d\alpha^n} \right|_{\alpha=0} = H_n(M, A)$$

is a martingale. For the proof that $H_n(M_t, A_t) \in L^2$ for each t, note that for each $m \in I\!\!N$,

$$E(|M_t|^m) \le K_m E\left(\exp(\alpha_0 |M_t|)\right) < \infty,$$

by assumption (b), and A_t is bounded, by assumption (a). Hence, since $H_n(M_t, A_t)$ is a polynomial function of M_t and A_t, it follows that $H_n(M_t, A_t)$ is in L^2 for each t and n. ∎

The polynomials $H_n(x, y)$ are related to the Hermite polynomials $h_n(x)$ by the formula

(6.12) $$H_n(x, y) = \left(\frac{y}{2}\right)^{\frac{1}{2}n} h_n\left(x/\sqrt{2y}\right).$$

For example, $H_0(x, y) = 1, H_1(x, y) = x, H_2(x, y) = x^2 - y, H_3(x, y) = x^3 - 3xy, H_4(x, y) = x^4 - 6x^2y + 3y^2$. From the known recursive formula (see Coddington [21, p. 131]):

$$\frac{d}{dz} h_n(z) = 2n h_{n-1}(z) \quad \text{for } n \in I\!\!N,$$

it follows that for each $n \in \mathbb{N}$,

(6.13) $$\frac{\partial}{\partial x} H_n(x,y) = n H_{n-1}(x,y) \quad \text{and}$$

(6.14) $$\frac{\partial}{\partial y} H_n(x,y) = \frac{n}{2y} H_n(x,y) - \frac{nx}{2y} H_{n-1}(x,y).$$

By using the recursive formula (from [21]):

$$h_n(z) = 2z h_{n-1}(z) - 2(n-1) h_{n-2}(z) \quad \text{for } n = 2,3,\ldots$$

or in terms of H_n:

$$H_n(x,y) = x H_{n-1}(x,y) - (n-1) y H_{n-2}(x,y) \quad \text{for } n = 2,3,\ldots,$$

we can simplify (6.14) to

(6.15) $$\frac{\partial}{\partial y} H_n(x,y) = -\frac{n(n-1)}{2} H_{n-2}(x,y) \quad \text{for } n = 2,3,\ldots.$$

Thus by (6.13) and (6.15) we have

(6.16) $$\frac{\partial}{\partial y} H_n(x,y) + \frac{1}{2} \frac{\partial^2}{\partial x^2} H_n(x,y) = 0 \quad \text{for } n = 2,3,\ldots.$$

By the forms of H_0 and H_1, this is seen to also hold for $n = 1, 2$.

Under the conditions of Theorem 6.4, when $M_0 = 0$ we have by Itô's formula:

$$H_n(M_t, A_t) - H_n(0,0) = \int_0^t \frac{\partial}{\partial x} H_n(M_s, A_s)\, dM_s + \int_0^t \frac{\partial}{\partial y} H_n(M_s, A_s)\, dA_s$$
$$+ \int_0^t \frac{1}{2} \frac{\partial^2}{\partial x^2} H_n(M_s, A_s)\, dA_s.$$

This reduces, by (6.13), (6.16), and the fact that $H_n(0,0) = 0$ for $n \geq 1$, to the following for $n = 1, 2, 3, \ldots,$

$$H_n(M_t, A_t) = \int_0^t n H_{n-1}(M_s, A_s)\, dM_s.$$

Since $H_0(M_t, A_t) = 1$, it then follows by induction that for each $n \in \mathbb{N}$,

$$(6.17) \qquad H_n(M_t, A_t) = n! \int\limits_0^t \int\limits_0^{t_{n-1}} \cdots \int\limits_0^{t_1} dM_s \, dM_{t_1} \ldots dM_{t_{n-1}}.$$

Thus we have the following complement to Theorem 6.4.

Theorem 6.5. *Suppose the conditions of Theorem 6.4 hold and $M_0 = 0$. Then for each $n \in \mathbb{N}$, the L^2-martingale $H_n(M, A)$ is given by a repeated stochastic integral, i.e., $H_n(M_t, A_t)$ is given by (6.17).*

We now illustrate the above results when $M_t = \int_0^t X_s \, dB_s$, where B is a Brownian motion in $\mathbb{R}$ and X is a bounded $\mathcal{B} \times \mathcal{F}$-measurable adapted process. By the results of Chapter 3 (Theorem 3.7 ff.), such an X is $\lambda \times P$-a.e. equal to a bounded predictable process with the same stochastic integral. Thus it suffices to consider X bounded and predictable.

Example. Let B be a Brownian motion in $\mathbb{R}$. As usual, let $\{\mathcal{F}_t, t \in \mathbb{R}_+\}$ be the filtration associated with B, and let $\mathcal{P}$ denote the class of predictable sets. Since $dB = d(B - B_0)$, we may suppose $B_0 = 0$. Let X be a bounded predictable process and suppose $X^2 \le C$, where C is a constant. Then $X \in \Lambda^2(\mathcal{P}, B)$ and hence $M = \{\int_0^t X_s \, dB_s, t \in \mathbb{R}_+\}$ is a continuous L^2-martingale. By (5.21) and the fact that $[B]_s = s$, it follows that the quadratic variation of M is given by

$$[M]_t = \int\limits_0^t (X_s)^2 \, ds.$$

Since X is bounded, $[M]_t \le Ct$ for all t. By Theorem 6.2, for each $\alpha \in \mathbb{R}$, $Z^\alpha = \exp(\alpha M - \frac{1}{2}\alpha^2[M])$ is a local martingale. Since Z^α is positive, it follows by Fatou's lemma that if $\{\tau_k\}$ is a localizing sequence for Z^α then

$$E(Z_t^\alpha) = E\left(\lim_{k \to \infty} Z_{t \wedge \tau_k}^\alpha\right) \le \liminf_{k \to \infty} E\left(Z_{t \wedge \tau_k}^\alpha\right)$$

where $E\left(Z^{\alpha}_{t\wedge T_k}\right) = E(Z^{\alpha}_0) = 1$. Combining this with $[M]_t \leq Ct$, we obtain

(6.18)
$$E\left[\exp\left(\alpha M_t\right)\right] = E\left[\exp\left(\frac{1}{2}\alpha^2[M]_t\right) Z^{\alpha}_t\right]$$
$$\leq \exp\left(\frac{1}{2}\alpha^2 Ct\right) E\left(Z^{\alpha}_t\right) \leq \exp\left(\frac{1}{2}\alpha^2 Ct\right).$$

Next consider

$$E\left(\int_0^t (Z^{\alpha}_s)^2\, d[M]_s\right) = E\left(\int_0^t \left(\exp\left(2\alpha M_s - \alpha^2[M]_s\right)\right)(X_s)^2\, ds\right)$$
$$\leq CE\left(\int_0^t \exp\left(2\alpha M_s\right)\, ds\right).$$

By applying Fubini's theorem in the last line above and using (6.18) with α replaced by 2α we obtain

$$E\left(\int_0^t (Z^{\alpha}_s)^2\, d[M]_s\right) \leq Ct \exp\left(2\alpha^2 Ct\right).$$

Thus condition (6.6) of Theorem 6.2 is satisfied and since $Z^{\alpha}_0 = 1$, it follows that Z^{α} is an L^2-martingale. By (6.18), for any $\alpha_0 > 0$ we have

(6.19)
$$E\left(\exp(\alpha_0 |M_t|)\right) \leq E\left(\exp(\alpha_0 M_t) + \exp(-\alpha_0 M_t)\right)$$
$$\leq 2\exp\left(\frac{1}{2}\alpha_0^2 Ct\right).$$

Thus the hypotheses of Theorem 6.4 are satisfied and consequently $H_n(M, [M])$ is an L^2-martingale for each $n \in \mathbb{N}_0$. By Theorem 6.5 we have the following representation for each $n \in \mathbb{N}$:

$$H_n(M_t, [M]_t) = n! \int_0^t \int_0^{t_{n-1}} \cdots \left(\int_0^{t_1} X_s\, dB_s\right) X_{t_1}\, dB_{t_1}\ldots X_{t_{n-1}}\, dB_{t_{n-1}}.$$

In summary, we have shown for any bounded predictable process X that $M = \{\int_0^t X_s\, dB_s, t \in \mathbb{R}_+\}$, $Z^{\alpha} = \exp(\alpha M - \frac{1}{2}\alpha^2[M])$, and $H_n(M, [M])$, are continuous L^2-martingales for each $\alpha \in \mathbb{R}$ and $n \in \mathbb{N}_0$. As an application of these results, we give an example for $n = 4$ of how one can obtain bounds for the moments of M using the fact that $H_n(M, [M])$ is a martingale.

For $n = 4$ we have

$$H_4(M_t, [M]_t) = (M_t)^4 - 6(M_t)^2[M]_t + 3([M]_t)^2$$

and by taking expectations we obtain

$$0 = E\left\{(M_t)^4\right\} - 6E\left\{(M_t)^2[M]_t\right\} + 3E\left\{([M]_t)^2\right\}.$$

Thus,

$$E\left\{(M_t)^4\right\} \leq 6E\left\{(M_t)^2[M]_t\right\} \leq 6\left(E\left\{(M_t)^4\right\} E\left\{([M]_t)^2\right\}\right)^{\frac{1}{2}}$$

where we have used Cauchy-Schwarz to obtain the second inequality. By squaring both sides and dividing by $E\{(M_t)^4\}$ (when it is non-zero), we obtain

$$E\left\{(M_t)^4\right\} \leq 36E\left\{([M]_t)^2\right\}.$$

Hence,

$$(6.20) \qquad E\left\{\left(\int_0^t X_s\, dB_s\right)^4\right\} \leq 36E\left\{\left(\int_0^t (X_s)^2\, ds\right)^2\right\} \leq 36C^2 t^2,$$

when $X^2 \leq C$. This inequality will be used in the next chapter.

6.4 Feynman-Kac Functional and the Schrödinger Equation

In this section, we apply the Itô formula to obtain a probabilistic representation for solutions of the (reduced) Schrödinger equation.

Let D be a bounded domain in $\mathbb{R}^d, d \geq 1$, with boundary ∂D and closure $\overline{D} = D \cup \partial D$. Let q be a bounded Borel measurable function on $\mathbb{R}^d$ that is zero outside of D. We write $E \subset\subset D$ to mean $E \subset \overline{E} \subset D$. We use $C^2(D)$ to denote the class of twice continuously differentiable functions in D and $C(\overline{D})$ to denote the class of continuous functions on $\overline{D}$. Let Δ

denote the Laplacian in $\mathbb{R}^d$ and m denote Lebesgue measure on the Borel sets in $\mathbb{R}^d$. We shall say that ψ is a solution of the Schrödinger equation in D if $\psi \in C^2(D)$ and

$$(6.21) \qquad \frac{1}{2}\Delta\psi + q\psi = 0 \quad m\text{-a.e. in } D.$$

Let $\mathbf{B}$ denote a Brownian motion in $\mathbb{R}^d$, let $\{\mathcal{F}_t, t \in \mathbb{R}_+\}$ be the usual filtration associated with $\mathbf{B}$, and let E^x denote the expectation given $\mathbf{B}_0 = x$. For each domain E in $\mathbb{R}^d$, let $\tau_E = \inf\{t > 0 : \mathbf{B}_t \notin E\}$, the first exit time of $\mathbf{B}$ from E. The probabilistic representation of solutions of the Schrödinger equation uses the Feynman-Kac functional which is defined by

$$e(t) = \exp\left(\int_0^t q(\mathbf{B}_s)\,ds\right) \qquad \text{for all } t \geq 0.$$

The following lemma gives a preliminary representation for solutions of (6.21). It is obtained by applying the multi-dimensional Itô formula to the processes $e(\cdot \wedge \tau_E)$ and $\mathbf{B}(\cdot \wedge \tau_E)$ and the function $(x, y) \to y\psi(x)$, where ψ is a solution of the Schrödinger equation in D and E is a domain such that $E \subset\subset D$.

Lemma 6.6. *Let ψ be a solution of the Schrödinger equation in D. Then for any domain $E \subset\subset D$, for each $x \in D$ and $t \in \mathbb{R}_+$ we have*

$$(6.22) \qquad \psi(x) = E^x\left\{e(t \wedge \tau_E)\psi(\mathbf{B}(t \wedge \tau_E))\right\}.$$

Proof. Let E be a domain such that $E \subset\subset D$. If $\mathbf{B}_0 = x \in D\backslash\overline{E}$, then $\tau_E = 0$ and (6.22) is trivially verified. Suppose $\mathbf{B}_0 = x \in \overline{E}$. Then for each $i \in \{1, \ldots, d\}$, $M^i = \{B^i_{t \wedge \tau_E}, t \in \mathbb{R}_+\}$ is a continuous L^2-martingale. Furthermore, $V = \{e(t \wedge \tau_E), t \in \mathbb{R}_+\}$ is a continuous process which is locally of bounded variation and

$$dV_t = \begin{cases} q(\mathbf{B}_t)e(t)dt & \text{for } 0 \leq t < \tau_E \\ 0 & \text{for } t \geq \tau_E. \end{cases}$$

For each t, $Z_t \equiv (M_t, V_t) \in \overline{E} \times \mathbb{R} \subset D \times \mathbb{R}$. The function $f : D \times \mathbb{R} \to \mathbb{R}$ defined by $f(x, y) = y\psi(x)$ for $x \in D$ and $y \in \mathbb{R}$ is continuous with continuous partial derivatives

$$\frac{\partial f}{\partial x_i}(x, y) = y \frac{\partial \psi}{\partial x_i}(x),$$

$$\frac{\partial^2 f}{\partial x_i \partial x_j}(x, y) = y \frac{\partial^2 \psi}{\partial x_i \partial x_j}(x),$$

$$\frac{\partial f}{\partial y}(x, y) = \psi(x)$$

on $D \times \mathbb{R}$ for $1 \leq i, j \leq d$. Thus the hypotheses of Theorem 5.10 are satisfied and we can apply the Itô formula to obtain a.s. for all t:

$$f(M_t, V_t) - f(M_0, V_0)$$

(6.23)
$$= \sum_{i=1}^{d} \int_0^t V_s \frac{\partial \psi}{\partial x_i}(M_s) M_s^i + \int_0^t \psi(M_s)\, dV_s$$

$$+ \frac{1}{2} \sum_{i=1}^{d} \sum_{j=1}^{d} \int_0^t V_s \frac{\partial^2 \psi}{\partial x_i \partial x_j}(M_s)\, d[M^i, M^j]_s.$$

By Corollary 5.4 we have $[M^i, M^j]_t = [B^i, B^j]_{t \wedge \tau_E}$, which equals $t \wedge \tau_E$ when $i = j$ and zero when $i \neq j$. Thus substituting for f, M, and V, in (6.23) yields

$$e(t \wedge \tau_E)\psi\left(\mathbf{B}(t \wedge \tau_E)\right) - \psi(\mathbf{B}_0)$$

(6.24)
$$= \sum_{i=1}^{d} \int_0^{t \wedge \tau_E} e(s) \frac{\partial \psi}{\partial x_i}(\mathbf{B}_s)\, dB_s^i + \int_0^{t \wedge \tau_E} \psi(\mathbf{B}_s) q(\mathbf{B}_s) e(s)\, ds$$

$$+ \frac{1}{2} \sum_{i=1}^{d} \int_0^{t \wedge \tau_E} e(s) \frac{\partial^2 \psi}{\partial x_i^2}(\mathbf{B}_s)\, ds.$$

Now, the Borel set of zero Lebesgue measure where equality in (6.21) does not hold is of zero potential for Brownian motion (cf. Chung [12, p. 112]). Consequently, P^x-a.s., the amount of time that $\mathbf{B}$ spends in this set has zero one-dimensional Lebesgue measure and hence the sum of the last two integrals in (6.24) is zero. Furthermore, for each i, $\frac{\partial \psi}{\partial x_i}$ is bounded on $\overline{E}$ and

$$e(s) \leq \exp\left(t \|q\|_{\overline{E}}\right) \quad \text{for} \quad 0 \leq s \leq t \wedge \tau_E$$

where $\|q\|_{\overline{E}} = \max_{y \in \overline{E}} |q(y)|$; hence $1_{[0,t \wedge \tau_E]}(s)e(s)\frac{\partial \psi}{\partial x_i}(\mathbf{B}_s)$ is bounded. Therefore, by the isometry (equation (2.10)), each term in the sum of stochastic integrals in (6.24) is an L^2-martingale. Thus by taking expectations in (6.24) we obtain

$$E^x \{e(t \wedge \tau_E)\psi(\mathbf{B}(t \wedge \tau_E))\} - \psi(x) = 0. \quad \blacksquare$$

We would like to let $t \to \infty$ in (6.22) and deduce that for all $x \in D$,

$$(6.25) \qquad \psi(x) = E^x \{e(\tau_E)\psi(\mathbf{B}_{\tau_E})\}.$$

However, to justify the interchange of limit and expectation we need to know that $\{e(t \wedge \tau_E)\psi(\mathbf{B}(t \wedge \tau_E)), t \in I\!\!R_+\}$ is uniformly integrable. A sufficient condition for this is that $E^x \{e(\tau_E)\} < \infty$ for some $x \in E$. Furthermore, if a solution ψ of the Schrödinger equation in D is continuous on $\overline{D}$, it is natural to ask whether by letting $t \to \infty$ and $E \uparrow D$ in (6.22), we can obtain the following representation for all $x \in D$:

$$(6.26) \qquad \psi(x) = E^x \{e(\tau_D)\psi(\mathbf{B}_{\tau_D})\}.$$

This can be established if $E^x \{e(\tau_D)\} < \infty$ for some $x \in D$. We shall prove this and as a corollary deduce the result pertaining to (6.25). We use the following result of Chung and Rao [18; Theorems 1.2 and 2.1, Lemma D], which we state without proof.

Proposition 6.7. Let $u_D(x) = E^x \{e(\tau_D)\}$ for all x in $\overline{D}$. If $u_D(x) < \infty$ for some $x \in D$, then u_D on $\overline{D}$ is bounded above and below by positive constants, and it is once continuously differentiable in D.

We only need the consequence of this theorem that if $u_D \not\equiv \infty$ in D, then u_D is finite and bounded below by a positive constant on $\overline{D}$. Some conditions which imply $u_D \not\equiv \infty$ in D are given at the end of this section. The function u_D is called the "gauge" for (D, q).

Theorem 6.8. Let ψ be a solution of the Schrödinger equation in D and suppose that $\psi \in C(\overline{D})$. Further suppose that $u_D(x) \equiv E^x \{e(\tau_D)\} < \infty$ for some $x \in D$. Then (6.26) holds for all x in $\overline{D}$.

Remark 1. This result shows that for a bounded domain D such that $u_D(x) < \infty$ for some $x \in D$, any solution of the Schrödinger equation in D taking given boundary values continuously on ∂D is unique and has the representation given by (6.26) for all $x \in \overline{D}$. Conversely, one can ask whether, given a continuous function ψ on ∂D, the right member of (6.26) defines a solution of the Schrödinger equation in D when $u_D(x) < \infty$ for some $x \in D$. To ensure sufficient differentiability of this candidate solution, some regularity of q is needed. For example, suppose q is locally Hölder continuous on D, i.e., for each compact set K in D there is $\alpha \in (0,1]$ such that

$$\sup_{x,y \in K, x \neq y} \frac{|q(x) - q(y)|}{|x - y|^{\alpha}} < \infty.$$

Then given a continuous function ψ on ∂D and assuming $u_D(x) < \infty$ for some $x \in D$, Chung and Rao [18, Theorems 1.3 and 2.1] have shown that the right member of (6.26) defines a solution of (6.21) everywhere on D. If we further assume that D is regular, i.e., $P^x\{\tau_D = 0\} = 1$ for all $x \in \partial D$, then this solution is continuous in $\overline{D}$ and agrees with ψ on ∂D. We shall not pursue the question of existence of solutions of (6.21) further here, since our focus is on representing given solutions, as an application of the Itô formula.

Remark 2. When $q \equiv 0$, the gauge is automatically finite, and (6.26) reduces to the usual probabilistic representation for solutions of the Dirichlet problem (see Chung [12, §4.4]).

Proof. Suppose $x \in D$ and let $\{E_n, n \in I\!N\}$ be a sequence of domains such that $x \in E_n \subset E_{n+1} \subset\subset D$ for all n and $\bigcup_n E_n = D$. By Proposition 6.7, $u_D(x) < \infty$ and

$$m_D \equiv \inf_{y \in \overline{D}} u_D(y) > 0.$$

By the strong Markov property of **B**, we have P^x-a.s.:

$$
\begin{aligned}
E^x\left\{e(\tau_D) \mid \mathcal{F}_{t \wedge \tau_{E_n}}\right\} &= e(t \wedge \tau_{E_n})E^{\mathbf{B}(t \wedge \tau_{E_n})}\{e(\tau_D)\} \\
(6.27) \qquad &= e(t \wedge \tau_{E_n})u_D\left(\mathbf{B}(t \wedge \tau_{E_n})\right) \\
&\geq e(t \wedge \tau_{E_n})m_D.
\end{aligned}
$$

Since $E^x\{e(\tau_D)\} = u_D(x) < \infty$, it follows that the left member above is uniformly integrable over all t and n, and hence so is $e(t \wedge \tau_{E_n})$ since $m_D > 0$. Moreover, ψ is bounded on $\overline{D}$ being continuous there. Hence, $\{e(t \wedge \tau_{E_n})\psi\left(\mathbf{B}(t \wedge \tau_{E_n})\right); t \in I\!\!R_+, n \in I\!\!N\}$ is uniformly integrable. By Lemma 6.6, for each t and n,

$$\psi(x) = E^x \left\{ e(t \wedge \tau_{E_n})\psi\left(\mathbf{B}(t \wedge \tau_{E_n})\right)\right\}.$$

Let $t \to \infty$ and then $n \to \infty$ in the above. Then (6.26) follows by uniformly integrability, the continuity of ψ on $\overline{D}$, and the fact that $\tau_{E_n} \uparrow \tau_D$ P^x-a.s. for all $x \in D$.

For $x \in \partial D$, let $\delta > 0$ and define

$$\tau_\delta = \inf\{t > 0 : |\mathbf{B}_t - \mathbf{B}_0| \geq \delta\}.$$

By the strong Markov property of $\mathbf{B}$ we have

$E^x\{e(\tau_D)\psi(\mathbf{B}_{\tau_D})\}$
$= E^x\{e(\tau_\delta)E^{\mathbf{B}_{\tau_\delta}}[e(\tau_D)\psi(\mathbf{B}_{\tau_D})]; \tau_\delta < \tau_D\} + E^x\{e(\tau_D)\psi(\mathbf{B}_{\tau_D}); \tau_D \leq \tau_\delta\}.$

By the representation already proved for $x \in D$, the expectation with respect to $P^{\mathbf{B}_{\tau_\delta}}$ above can be replaced by $\psi(\mathbf{B}_{\tau_\delta})$. In addition, since ψ is continuous on $\overline{D}$, given $\varepsilon > 0$ there is $\delta = \delta(\varepsilon) > 0$ such that $|\psi(x) - \psi(y)| < \varepsilon$ whenever $|x - y| \leq \delta$. Then it follows from the above that

$$|\psi(x) - E^x\{e(\tau_D)\psi(\mathbf{B}_{\tau_D})\}| \leq \varepsilon(E^x\{e(\tau_\delta); \tau_\delta < \tau_D\} + E^x\{e(\tau_D); \tau_D \leq \tau_\delta\})$$
$$= \varepsilon E^x\{e(\tau_\delta \wedge \tau_D)\}.$$

In a similar manner to (6.27) we have

$$E^x\{e(\tau_D)|\mathcal{F}_{\tau_\delta \wedge \tau_D}\} = e(\tau_\delta \wedge \tau_D)E^{\mathbf{B}_{\tau_\delta \wedge \tau_D}}\{e(\tau_D)\} \geq e(\tau_\delta \wedge \tau_D)m_D,$$

and consequently,

$$E^x\{e(\tau_\delta \wedge \tau_D)\} \leq E^x\{e(\tau_D)\}/m_D.$$

Since the right member above is independent of δ, and $\varepsilon > 0$ was arbitrary, it follows that (6.26) holds for $x \in \partial D$. ∎

Remark. If D is regular, the proof that (6.26) holds for $x \in \partial D$ is trivial, because then $P^x(\tau_D = 0) = 1$ for $x \in \partial D$. The assumed continuity of the solution ψ on $\overline{D}$ allows us to prove (6.26) holds on ∂D without any additional assumption on ∂D.

If ψ is a solution of the Schrödinger equation in D, but ψ is not necessarily in $C(\overline{D})$, we can obtain a representation theorem for ψ on any domain $E \subset\subset D$ as follows, whenever the gauge for E is finite.

Corollary 6.9. *Let ψ be a solution of the Schrödinger equation in D. Let E be a domain such that $E \subset\subset D$. Suppose that*

$$u_E(x) = E^x\{e(\tau_E)\} < \infty \quad \text{for some } x \in E.$$

Then (6.25) holds for all x in D.

Proof. Since ψ is twice continuously differentiable in a domain containing $\overline{E}$, it is continuous on $\overline{E}$ and so for $x \in \overline{E}$, (6.25) follows immediately from Theorem 6.8 applied to $\overline{E}$. For $x \in D \backslash \overline{E}$, $\tau_E = 0$ P^x-a.s. and (6.25) holds trivially. ∎

Remark. A sufficient condition for $u_E(x) < \infty$ for some $x \in E$ is that $u_D(x) < \infty$ for some $x \in D$ (see Exercises).

To conclude this discussion of the probabilistic representation of solutions of the Schrödinger equation in D, we mention some conditions under which the gauge $u_D(x) = E^x\{e(\tau_D)\}$ is finite. The first of these is the following.

Theorem 6.10. *Suppose there is a solution $\phi \in C(\overline{D})$ of the Schrödinger equation in D such that $\phi > 0$ in $\overline{D}$. Then $u_D(x) < \infty$ for all $x \in D$.*

Proof. By Lemma 6.6, for any domain $E \subset\subset D, x \in D$, and $t \in \mathbb{R}_+$,

$$\phi(x) = E^x\{e(t \wedge \tau_E)\phi(\mathbf{B}(t \wedge \tau_E))\} \geq E^x\{e(t \wedge \tau_E)\} c_D$$

where $c_D \equiv \min_{y \in \overline{D}} \phi(y) > 0$. Letting $t \to \infty$ and then $E \uparrow D$ it follows by Fatou's lemma that

$$u_D(x) = E^x \{e(\tau_D)\} \leq \frac{\phi(x)}{c_D} < \infty. \quad \blacksquare$$

It follows from Remark 1 following Theorem 6.8, and Theorem 6.10, that for q locally Hölder continuous in D and D a regular domain, $u_D \not\equiv \infty$ in D is equivalent to the existence of a solution $\phi \in C(\overline{D})$ of the Schrödinger equation in D such that $\phi > 0$ in $\overline{D}$. In [78], Williams showed that if the boundary of D is C^2 (see [35, Section 6.2] for the definition) and $E^x \{e(\tau_D)1_A(\mathbf{B}_{\tau_D})\}$ is finite for some $x \in D$ and some non-empty relatively open subset A of ∂D, then $u_D(x) < \infty$. Subsequent results of Falkner [31] imply that this remains true if A simply has positive boundary measure. In [17], Chung, Li and Williams made the connection between the finiteness of u_D and the signs of the eigenvalues of the operator $L = \frac{1}{2}\Delta + q$. A real number α is an eigenvalue of the operator L on D if there is $\psi \in C^2(D) \cap C(\overline{D}), \psi \not\equiv 0$, such that

$$L\psi = \alpha\psi \text{ in } D \text{ and } \psi = 0 \text{ on } \partial D.$$

It follows from [17] that when q is locally Hölder continuous in D and D is regular, the following three propositions are equivalent:

(i) $u_D(x) < \infty$ for some $x \in D$,

(ii) there is no non-negative eigenvalue for the operator L on D,

(iii) there is a solution ϕ of the Schrödinger equation in D such that $\phi \in C(\overline{D})$ and $\phi > 0$ in $\overline{D}$.

Generalizations of the above results on the relationship between finiteness of the gauge and validity of (6.26) as a representation for solutions of the Schrödinger equation have been obtained by a variety of authors in recent years. These include extensions to the case of unbounded potentials q [1, 83] or where q is replaced by a measure [6], unbounded domains [20], conditioned Brownian motion [32, 83, 84, 22], other diffusions and Markov processes [22, 14], and other boundary conditions of Neumann or mixed type [16, 44, 61]. For a survey of some of these, the reader is referred to

[13] or the monograph [20]. As an illustration, we state a result for the case of unbounded q. Details may be found in [20]. For this, we replace the assumption that q is bounded on D with the assumption that q satisfies the following. Recall that $q \equiv 0$ outside of D.

$$\lim_{t \to 0} \sup_{x \in I\!R^d} E^x \left[\int_0^t |q(\mathbf{B}_s)|\, ds \right] = 0.$$

It can be shown that this is equivalent to the analytic condition:

$$\lim_{\alpha \to 0} \sup_{x \in I\!R^d} \int_{|x-y| \le \alpha} G(x, y)\, |q(y)|\, dy = 0,$$

where $G(x, y) = |x - y|^{2-d}$ if $d \ge 3$, $G(x, y) = \max(-\log|x - y|, 0)$ if $d = 2$ and $G(x, y) = |x - y|$ if $d = 1$. We say q is of Kato class and write $q \in K_d$. It follows from this condition that

$$0 < e(t) \equiv \exp\left(\int_0^t q(\mathbf{B}_s)\, ds \right) < \infty \quad P^x\text{-a.s. for all } x \in I\!R^d,$$

and we have the following theorem.

Theorem 6.11. *Suppose $q \in K_d$. Then either $u_D \equiv \infty$ in D or u_D is bounded above and below by positive constants in D and u_D is continuous in D. Moreover, in the latter case, for any $\psi \in C(\overline{D})$ that satisfies (6.21) in the sense of Schwartz distributions, we have*

$$(6.28) \qquad \psi(x) = E^x \left[e(\tau_D)\psi(\mathbf{B}(\tau_D)) \right] \qquad \text{for all } x \in \overline{D}.$$

6.5 Exercises

1. Let $M = \{M_t, t \ge 0\}$ be a continuous real-valued process that is adapted to a given standard filtration $\{\mathcal{F}_t\}$. Taking all (local) martingales to be defined relative to this filtration, prove that (i)–(v) below are equivalent. Note this result yields several alternative forms for the characterization of a one-dimensional Brownian given in Theorem 6.1.

(i) $\{\exp(\theta M_t - \frac{1}{2}\theta^2 t), t \ge 0\}$ is a local martingale for each $\theta \in I\!R$.

(ii) M and $\{M_t^2 - t, t \geq 0\}$ are local martingales.

(iii) M is a local martingale with quadratic variation process: $[M]_t = t$ a.s. for all $t \geq 0$.

(iv) For each continuous real-valued function f that has continuous first and second partial derivatives on $\mathbb{R}$, $\{f(M_t) - \frac{1}{2}\int_0^t f''(M_s)ds, t \geq 0\}$ is a local martingale.

(v) $\{\exp(i\theta M_t + \frac{1}{2}\theta^2 t), t \geq 0\}$ is a complex-valued martingale for each $\theta \in \mathbb{R}$.

2. (a) Prove the following multi-dimensional version of Theorem 6.1. A d-dimensional process $M = \{M_t, t \in \mathbb{R}_+\}$ is a Brownian motion in $\mathbb{R}^d$ if and only if there is a standard filtration $\{\mathcal{F}_t\}$ such that M^i, $i = 1, \ldots, d$ are continuous local martingales relative to $\{\mathcal{F}_t\}$ and for each $i, j \in \{1, \ldots, d\}$,

$$[M^i, M^j]_t = \delta_{ij}t \quad \text{a.s. for each } t \geq 0,$$

where δ_{ij} denotes the Kronecker delta, defined to equal 1 if $i = j$ and to equal 0 otherwise.

Hint for the "if" part: For each $\theta \in \mathbb{R}^d$ and $r, s \geq 0$, show that:

$$E[\exp(i\theta \cdot (M_{s+r} - M_s))|\mathcal{F}_s] = \exp\left(-\frac{1}{2}|\theta|^2 r\right),$$

where $|\cdot|$ denotes the Euclidean norm on $\mathbb{R}^d$.

(b) Now generalize Exercise 1 to give several alternative forms of the local martingale characterization in 2(a) above.

In the next three exercises, B denotes a one-dimensional Brownian motion starting from $x \in \mathbb{R}$, and P^x and E^x denote the associated probability and expectation, respectively.

3. Use Theorem 6.2 to verify that under P^x,

$$\exp\left(\alpha B_t - \frac{1}{2}\alpha^2 t\right), t \geq 0,$$

defines an L^2-martingale for each $\alpha \in \mathbb{R}$.

4. Fix $a \in \mathbb{R}$ and define $\tau_a = \inf\{t \geq 0 : B_t = a\}$. Use the result of Exercise 3 above and Doob's stopping theorem (Corollary 1.7) to compute the value of the Laplace transform $E^x[\exp(-\theta\tau_a)]$ for each $\theta > 0$ and $x \in \mathbb{R}$.

Hint: Let $\theta = \alpha^2$ for a suitable choice of the sign of α depending on whether $x \leq a$ or $x > a$.

This Laplace transform can be inverted to yield the distribution of τ_a:

$$P^x(\tau_a \in dt) = \frac{|x - a|}{\sqrt{2\pi t^3}} \exp\left\{\frac{-(x-a)^2}{2t}\right\} dt, \ t \geq 0.$$

5. Let $\gamma \in \mathbb{R}$ and $X_t = B_t + \gamma t$ for all $t \geq 0$. Then X is a Brownian motion with constant drift γ. Define $\tau_0 = \inf\{t \geq 0 : X_t = 0\}$. Depending on the signs of γ and x, τ_0 may take the value $+\infty$ with positive P^x-probability. Use the pair of linearly independent solutions of the ordinary differential equation:

$$\frac{1}{2}f'' + \gamma f' + \alpha f = 0$$

to determine the value of

$$E^x[\exp(\alpha\tau_0)]$$

for all $x \in \mathbb{R}$ and α sufficiently small.

Hint: It suffices by symmetry to consider $x > 0$.

6. Let **B** denote a d-dimensional Brownian motion. For each $x \in \mathbb{R}^d$, let P^x and E^x denote the probability and expectation, respectively, associated with **B** starting from x. Suppose D is a bounded domain in $\mathbb{R}^d$ and h is a continuous real-valued function on $\overline{D}$. Define $\tau_D = \inf\{t > 0 : \mathbf{B}_t \notin D\}$. Prove that the following three statements are equivalent.

(i) h is harmonic in D.

(ii) $\{h(\mathbf{B}(t \wedge \tau_D)), t \geq 0\}$ is a martingale under P^x for each $x \in D$.

(iii) $h(x) = E^x[h(\mathbf{B}(\tau_D))]$ for each $x \in D$.

Hint for (iii) $\Rightarrow$ (i): Use the sphere averaging characterization of a harmonic function (see Chung [12, p. 156]).

7. Verify the Remark following Corollary 6.9.

8. Consider the one-dimensional differential equation

$$(6.29) \qquad \frac{1}{2}u'' + \alpha u = 0 \quad \text{for all } x \in (0, \pi).$$

Using the explicit two-dimensional family of solutions of (6.29), verify that statements (ii) and (iii) in the paragraph following Theorem 6.10 are equivalent to each other and to $\alpha < \frac{1}{2}$. Note that by the Fredholm alternative theorem, for $\alpha \neq n^2/2$, $n = 1, 2, \ldots$, there is a unique solution of (6.29) with any given boundary values, but the representation (6.26) only holds for $\alpha < 1/2$.

7

LOCAL TIME AND
TANAKA'S FORMULA

7.1 Introduction

In this chapter B denotes a Brownian motion in $I\!R$. For each $x \in I\!R$ we shall obtain a decomposition, known as Tanaka's formula, of the positive submartingale $|B - x|$ as the sum of another Brownian motion $\hat{B}$ and a continuous increasing process $L(\cdot , x)$. The latter is called the local time of B at x, a fundamental notion invented by P. Lévy (see [54]). It may be expressed as follows:

(7.1)
$$L(t, x) = \lim_{\varepsilon \downarrow 0} \frac{1}{2\varepsilon} \int_0^t 1_{(x-\varepsilon, x+\varepsilon)}(B_s) \, ds$$
$$= \lim_{\varepsilon \downarrow 0} \frac{1}{2\varepsilon} \lambda\{s \in [0, t] : B_s \in (x - \varepsilon, x + \varepsilon)\}$$

where λ is the Lebesgue measure. Thus it measures the amount of time the Brownian motion spends in the neighborhood of x. It is well known that $\{t \in I\!R_+ : B_t = x\}$ is a perfect closed set of Lebesgue measure zero. The existence of a nonvanishing L defined in (7.1) is therefore far from obvious. In fact, the limit in (7.1) exists both in L^2 and a.s., as we shall see. Moreover, $L(t, x)$ may be defined to be a jointly continuous function of (t, x). This was first proved by H. F. Trotter, but our approach follows that of Stroock and Varadhan [73, p. 117]. The local time plays an important role in many refined developments of the theory of Brownian motion. One application, given at the end of Section 7.3, is a derivation of the exponential distribution of the local time accumulated up until the hitting time of a fixed level. Other applications of local time and Tanaka's formula are discussed in the next two chapters.

Let us reveal here that Lévy's original proof of (7.1) was given before a theory of stochastic integration existed, and was based on a profound study of the structure of the zeros of $B(\cdot)$. For a historical perspective and a sketch of his method, see the exposition in Chung [15].

7.2 Local Time

For each $x \in \mathbb{R}$, define the function f_x by $f_x(y) = (y - x)^+$. This function is not differentiable but its first two derivatives in the sense of generalized functions (Schwartz distributions) are: $f'_x = 1_{[x,\infty)}$ and $f''_x = \delta_x$ where δ_x is the famous Dirac delta function. A formal application of Itô's formula yields:

$$(7.2) \qquad (B_t - x)^+ - (B_0 - x)^+ = \int_0^t 1_{[x,\infty)}(B_s)\,dB_s + \frac{1}{2}\int_0^t \delta_x(B_s)\,ds.$$

It turns out that this formula holds a.s. provided the last integral is interpreted as the limit in (7.1). We begin the proof by approximating f_x by $f_{x\varepsilon}$ ($\varepsilon > 0$), defined as follows:

$$f_{x\varepsilon}(y) = \begin{cases} 0, & \text{for } y \le x - \varepsilon \\ (y - x + \varepsilon)^2/4\varepsilon, & \text{for } x - \varepsilon \le y \le x + \varepsilon \\ y - x, & \text{for } y \ge x + \varepsilon \end{cases}$$

$$f'_{x\varepsilon}(y) = \begin{cases} 0, & \text{for } y \le x - \varepsilon \\ (y - x + \varepsilon)/2\varepsilon, & \text{for } x - \varepsilon \le y \le x + \varepsilon \\ 1, & \text{for } y \ge x + \varepsilon \end{cases}$$

$$f''_{x\varepsilon}(y) = \begin{cases} 0, & \text{for } y < x - \varepsilon \\ 1/2\varepsilon, & \text{for } x - \varepsilon < y < x + \varepsilon \\ 0, & \text{for } y > x + \varepsilon \end{cases}$$

Note that $f''_{x\varepsilon}$ is not defined at $x \pm \varepsilon$, but we set it to be zero there. The graphs of $f_{x\varepsilon}$ and its derivatives are sketched in Figures 7.1–7.3.

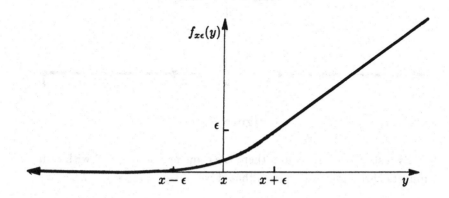

Figure 7.1.

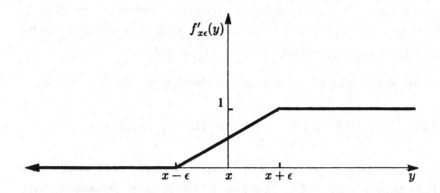

Figure 7.2.

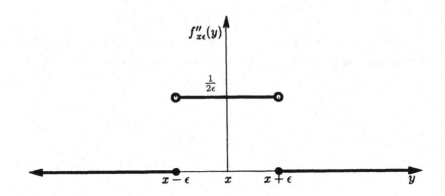

Figure 7.3.

By a standard procedure, there is a sequence of $\phi_n \in C^\infty$ with compact supports shrinking to $\{0\}$, such that if we put $g_n = \phi_n * f_{x\epsilon}$, i.e.,

$$g_n(y) = \int_{I\!\!R} f_{x\epsilon}(y - z)\phi_n(z)\,dz,$$

then $g_n \in C^\infty$ and $g_n \to f_{x\epsilon}$, $g_n' \to f_{x\epsilon}'$ both uniformly in $I\!\!R$, while $g_n'' \to f_{x\epsilon}''$ pointwise except at $x \pm \epsilon$. In fact, ϕ_n may be defined as follows: $\phi_n(y) = n\phi(ny)$, where $\phi(y) = c\exp\left(-(1-y^2)^{-1}\right)$ for $|y| < 1$ and $\phi(y) = 0$ for $|y| \geq 1$; the constant c being such that $c\int_{-1}^{1} \phi(y)\,dy = 1$.

By applying Itô's formula to g_n, we obtain almost surely:

$$(7.3) \qquad g_n(B_t) - g_n(B_0) = \int_0^t g_n'(B_s)\,dB_s + \frac{1}{2}\int_0^t g_n''(B_s)\,ds.$$

For each t, as $n \to \infty$, $1_{[0,t]}g_n'(B)$ converges to $1_{[0,t]}f_{x\epsilon}'(B)$ uniformly on $I\!\!R_+ \times \Omega$ and hence in $\mathcal{L}^2$. Therefore, by the isometry (equation (2.10)), $\int_0^t g_n'(B_s)\,dB_s$ converges in L^2 to $\int_0^t f_{x\epsilon}'(B_s)\,dB_s$. On the other hand, since $P(B_s = x \pm \epsilon) = 0$ for each s, $\lim_{n\to\infty} g_n''(B_s) = f_{x\epsilon}''(B_s)$ a.s. for each fixed s. Hence, by Fubini's theorem, this limit relation also holds for λ-almost all s in $\mathbb{R}_+$, almost surely. Since $|g_n''| \le (2\varepsilon)^{-1}$, it follows by bounded convergence that $\int_0^t g_n''(B_s)\,ds$ converges to $\int_0^t f_{x\varepsilon}''(B_s)\,ds$ both a.s. and in L^2. Thus, by letting $n \to \infty$ in (7.3), we obtain for each x and t, almost surely:

(7.4)
$$f_{x\varepsilon}(B_t) - f_{x\varepsilon}(B_0) = \int_0^t f_{x\varepsilon}'(B_s)\,dB_s$$
$$+ \frac{1}{2}\int_0^t \frac{1}{2\varepsilon} 1_{(x-\varepsilon,x+\varepsilon)}(B_s)\,ds.$$

As $\varepsilon \downarrow 0$ in (7.4), $f_{x\varepsilon}(B_t) - f_{x\varepsilon}(B_0)$ converges to $(B_t - x)^+ - (B_0 - x)^+$ in L^2, since $|f_{x\varepsilon}(B_t) - f_{x\varepsilon}(B_0)| \le |B_t - B_0|$. Also we have

$$E\left\{\int_0^t \left(f_{x\varepsilon}'(B_s) - 1_{[x,\infty)}(B_s)\right)^2\,ds\right\} \le E\left\{\int_0^t 1_{(x-\varepsilon,x+\varepsilon)}(B_s)\,ds\right\}$$
$$\le \int_0^t \frac{2\varepsilon}{\sqrt{2\pi s}}\,ds \to 0$$

as $\varepsilon \to 0$. Hence by the isometry (2.10), $\int_0^t f_{x\varepsilon}'(B_s)\,dB_s$ converges to $\int_0^t 1_{[x,\infty)}(B_s)\,dB_s$ in L^2. Thus the following has been proved.

Theorem 7.1. *For each $t \in \mathbb{R}_+$ and $x \in \mathbb{R}$, we have almost surely,*

(7.5)
$$(B_t - x)^+ - (B_0 - x)^+ = \int_0^t 1_{[x,\infty)}(B_s)\,dB_s + \frac{1}{2}L(t,x)$$

where L is defined in (7.1) with the limit there in L^2.

Recall from Chapter 2 that $\int_0^t 1_{[x,\infty)}(B_s)\,dB_s$ denotes a version of the stochastic integral which is continuous in t for each fixed x. Indeed, as a process indexed by t it is a continuous L^2-martingale. However, we may also regard it as a stochastic process indexed by (t,x). Viewed in this way, it turns out that there is a version of it which is continuous in (t,x). We state this as a lemma, but postpone its proof until later (see Section 7.4).

Remark. Strictly speaking, a stochastic process indexed by two parameters has not been formally defined. However, it is a direct generalization of the one-parameter case. For instance, $\{X(t, x), (t, x) \in \mathbb{R}_+ \times \mathbb{R}\}$ is a stochastic process iff $X(t, x)$ is a random variable for each (t, x).

Lemma 7.2. *There exists a family of random variables* $\{J(t, x), (t, x) \in \mathbb{R}_+ \times \mathbb{R}\}$ *and a set* Ω_0 *with* $P(\Omega_0) = 1$ *such that* $(t, x) \to J(t, x)(\omega)$ *is continuous for all* $\omega \in \Omega_0$ *and for each fixed* (t, x):

(7.6)
$$P \left\{ \int_0^t 1_{[x,\infty)}(B_s) \, dB_s = J(t, x) \right\} = 1.$$

By shrinking the sample space to Ω_0 we may and do assume that J is continuous for all ω. We now redefine L by means of (7.5), using J as the version of the stochastic integral there.

Definition. For each $(t, x) \in \mathbb{R}_+ \times \mathbb{R}$, let

(7.7)
$$\frac{1}{2} L(t, x) = (B_t - x)^+ - (B_0 - x)^+ - J(t, x).$$

Since we assume $t \to B_t$ to be continuous (for all ω), L is clearly continuous in (t, x), because J is. Owing to (7.6), for each fixed (t, x), (7.5) remains true with this new definition of L and so does (7.1).

The next result embodies a remarkable change of space and time integration, and exhibits the local time as the true density of occupation time.

Theorem 7.3. *For each* t *and* $a \leq b$, *we have almost surely,*

(7.8)
$$\int_a^b L(t, x) \, dx = \int_0^t 1_{(a,b)}(B_s) \, ds.$$

Proof. Since

$$f'_{x\varepsilon}(z) = \frac{1}{2\varepsilon} \int\limits_{x-\varepsilon}^{x+\varepsilon} 1_{[y,\infty)}(z)\, dy,$$

we have almost surely,

$$\int\limits_0^t f'_{x\varepsilon}(B_s)\, dB_s = \frac{1}{2\varepsilon} \int\limits_0^t \int\limits_{x-\varepsilon}^{x+\varepsilon} 1_{[y,\infty)}(B_s)\, dy\, dB_s$$

(7.9)
$$= \frac{1}{2\varepsilon} \int\limits_{x-\varepsilon}^{x+\varepsilon} \int\limits_0^t 1_{[y,\infty)}(B_s)\, dB_s\, dy$$

$$= \frac{1}{2\varepsilon} \int\limits_{x-\varepsilon}^{x+\varepsilon} J(t,y)\, dy$$

provided that the change of order of integrations is allowed. This interchange is not covered by the usual Fubini theorem since it involves a stochastic integral. To obtain the result rigorously, we approximate the last
integral by Riemann sums. Let

$$\phi_n(z) = \sum_k \frac{1}{2^n} 1_{[k2^{-n},\infty)}(z)$$

where the sum is over all k such that $k2^{-n} \in (x - \varepsilon, x + \varepsilon)$. Since $J(t, \,\cdot\,)$
is continuous, using (7.6) we obtain a.s.:

(7.10)
$$\int\limits_{x-\varepsilon}^{x+\varepsilon} J(t,y)\, dy = \lim_{n\to\infty} \int\limits_0^t \phi_n(B_s)\, dB_s.$$

As $n \to \infty$, ϕ_n converges uniformly to $2\varepsilon f'_{x\varepsilon}$, hence by the isometry (2.10),
the limit in (7.10) is equal a.s. to $2\varepsilon \int_0^t f'_{x\varepsilon}(B_s)\, dB_s$. This proves the almost
sure equality of the first and last members in (7.9), as desired.

By substituting this result into (7.4), we obtain a.s.

(7.11)
$$f_{x\varepsilon}(B_t) - f_{x\varepsilon}(B_0) - \frac{1}{2\varepsilon} \int\limits_{x-\varepsilon}^{x+\varepsilon} J(t,y)\,dy$$
$$= \frac{1}{2} \int\limits_0^t \frac{1}{2\varepsilon} 1_{(x-\varepsilon,x+\varepsilon)}(B_s)\,ds.$$

We want to integrate this over x. But since (7.11) holds a.s. for each x only, we must approximate again by Riemann sums so that only a countable set of x will be used, for which (7.11) holds simultaneously a.s. Let $F(x)$ denote the left member of (7.11). Since F is continuous, we have

$$\int\limits_a^b F(x)\,dx = \lim_{n\to\infty} \sum_k 2^{-n} F\left(k2^{-n}\right)$$

where the sum is over all k such that $k2^{-n} \in (a,b)$. By (7.11), this integral is equal a.s. to the limit of the following as $n \to \infty$:

$$\sum_k \frac{1}{2^n} \frac{1}{4\varepsilon} \int\limits_0^t 1_{(k2^{-n}-\varepsilon,k2^{-n}+\varepsilon)}(B_s)\,ds$$
$$= \frac{1}{4\varepsilon} \int\limits_0^t \left\{ \sum_k 2^{-n} 1_{(B_s-\varepsilon,B_s+\varepsilon)}\left(k2^{-n}\right) \right\} ds.$$

The sum in braces is dominated by $2\varepsilon + 1$ and converges as $n \to \infty$ to

$$\int\limits_a^b 1_{(B_s-\varepsilon,B_s+\varepsilon)}(x)\,dx.$$

It follows that a.s.:

(7.12)
$$\int\limits_a^b \left\{ f_{x\varepsilon}(B_t) - f_{x\varepsilon}(B_0) - \frac{1}{2\varepsilon} \int\limits_{x-\varepsilon}^{x+\varepsilon} J(t,y)\,dy \right\} dx$$
$$= \frac{1}{4\varepsilon} \int\limits_0^t \int\limits_a^b 1_{(x-\varepsilon,x+\varepsilon)}(B_s)\,dx\,ds.$$

Now for $z \in \mathbb{R}$,

(7.13) $\qquad \lim_{\varepsilon \downarrow 0} \frac{1}{2\varepsilon} \int_a^b 1_{(x-\varepsilon, x+\varepsilon)}(z)\, dx = 1_{(a,b)}(z) + \frac{1}{2}1_{\{a\}}(z) + \frac{1}{2}1_{\{b\}}(z).$

By letting $\varepsilon \downarrow 0$ in (7.12), using the continuity of $J(t, \cdot)$ and (7.13), we obtain a.s.

$$\int_a^b \{(B_t - x)^+ - (B_0 - x)^+ - J(t,x)\}\, dx = \frac{1}{2}\int_0^t 1_{(a,b)}(B_s)\, ds$$

since $1_{\{c\}}(B_s) = 0$ a.s., for each c and s. This is (7.8) in view of (7.7). ∎

The general form of Theorem 7.3 will be stated as a corollary. It is easily deduced from the theorem by the usual approximation or by a monotone class argument.

Corollary 7.4. *Let f be Borel measurable and locally integrable on $\mathbb{R}$, then we have for each t, almost surely:*

(7.14) $\qquad \displaystyle\int_{-\infty}^{\infty} L(t,x)f(x)\, dx = \int_0^t f(B_s)\, ds.$

We observe that the integral on the left side of (7.14) is actually over a bounded set of x, since for t and ω fixed, by (7.1) the set of x for which $L(t,x)(\omega) \neq 0$ is bounded, because the range of $B(s,\omega)$ for $s \in [0,t]$ is a compact set.

We can now prove that (7.1) holds also when the limits are taken in the almost sure sense. This follows from (7.11), because the left member is continuous in $\varepsilon > 0$, while the right member is left continuous. Hence for almost every ω, (7.11) holds simultaneously for all $\varepsilon > 0$. As $\varepsilon \downarrow 0$, the left member converges to $\frac{1}{2}L(t,x)$ by the continuity of $J(t, \cdot)$ and (7.7), proving the assertion.

7.3 Tanaka's Formula

Theorem 7.5. *For each (t, x), we have a.s.*

$$(7.15) \qquad |B_t - x| - |B_0 - x| = \int_0^t \mathrm{sgn}(B_s - x) \, dB_s + L(t, x).$$

Here $\mathrm{sgn}(y)$ is $1, 0,$ or -1, as y is greater, equal, or less than zero, respectively.

Proof. Since $-B$ is a Brownian motion, it has a local time at $-x$ which will be denoted by $L^-(t, -x)$. By applying (7.1) to it, we see at once that $L^-(t, -x) = L(t, x)$ a.s. By combining this with (7.5) applied to $-B$ and $-x$, instead of B and x, we obtain after some obvious sign switching:

$$(B_t - x)^- - (B_0 - x)^- = - \int_0^t 1_{(-\infty, x]}(B_s) \, dB_s + \frac{1}{2} L(t, x).$$

Adding this to (7.5), we obtain (7.15), since $\int_0^t 1_{\{x\}}(B_s) \, dB_s = 0$ a.s. ■

Equation (7.15) is known as *Tanaka's formula*. For each x it provides a decomposition of the submartingale $|B - x|$ into components which can be described more precisely than by the general form of the Doob-Meyer decomposition theorem.

Definition. For each t and x, let

$$(7.16) \qquad \hat{B}(t, x) = |B_0 - x| + \int_0^t \mathrm{sgn}(B_s - x) \, dB_s.$$

Theorem 7.6. *For each x, we have a.s.*

$$(7.17) \qquad |B - x| = \hat{B}(\,\cdot\,, x) + L(\,\cdot\,, x)$$

where $\hat{B}(\,\cdot\,, x)$ is a Brownian motion and $L(\,\cdot\,, x)$ is a continuous increasing process with initial value zero. Moreover, almost surely, $L(\,\cdot\,, x)$ can *increase only when* $|B - x|$ *is at zero, i.e.,*

$$\int_0^\infty 1_{\{t:B(t)\neq x\}} \, dL(t, x) = 0 \text{ a.s.}$$

Proof. Since both sides of (7.15) are continuous in t, (7.17) follows immediately. By (7.16), $\hat{B}(t, x)$ is the sum of its initial r.v. and the continuous L^2-martingale $\int_0^t \text{sgn}(B_s - x) \, dB_s$. The quadratic variation of the latter is $\int_0^t (\text{sgn}(B_s - x))^2 \, d[B]_s = t$ a.s., by Corollary 5.9. Thus, $\hat{B}(\cdot, x)$ is a continuous local martingale with quadratic variation process indistinguishable from $\{t, t \in \mathbb{R}_+\}$, hence it is a Brownian motion by Theorem 6.1.

The continuity and initial value of $L(\cdot, x)$ follow from (7.7). For each fixed t, $L(t, x)$ is given a.s. by the limit in (7.1), where the integral there is increasing in t for each fixed ε. It follows that $L(t, x)$ is a.s. increasing in t.

To prove the assertion about points of increase of $L(\cdot, x)$, we observe that there is a set Ω_1 such that $P(\Omega_1) = 1$ and (7.1) holds pointwise on Ω_1 simultaneously for all rational t. Suppose $\omega \in \Omega_1$ and t is a point of increase of $L(\cdot, x)(\omega)$. Then there are rationals $r < r'$ arbitrarily close to t such that $L(r, x)(\omega) < L(r', x)(\omega)$ and hence by (7.1),

$$\lambda\{s \in [r, r'] : B_s(\omega) \in (x - \varepsilon, x + \varepsilon)\} > 0 \quad \text{for all} \quad \varepsilon > 0.$$

Thus, by the continuity of $B.(\omega)$, we have $B_s(\omega) = x$ for some $s \in [r, r']$ and since r, r' were arbitrarily close to t it follows that $B_t(\omega) = x$. ∎

The particular case $x = 0$ of (7.17) will be discussed in more detail in the next chapter.

To illustrate some of the properties of local time developed in this chapter, we conclude this section with the following theorem. This result can be proved by bare-hands calculation as in Itô and McKean [47]. Here we adopt the streamlined "martingale proof" of Revuz and Yor [64].

Theorem 7.7. Suppose $B_0 = x$ almost surely, where $x \in \mathbb{R}$. Let $y \in \mathbb{R}$ and define $\tau_y = \inf\{t \geq 0 : B_t = y\}$. Then $L(\tau_y, x)$ is an exponential random variable with mean $2|y - x|$.

Proof. By the spatial homogeneity and symmetry of Brownian motion, it suffices to consider $x = 0$ and $y > 0$. Restricting to this case, we will establish the Laplace transform relation

$$E^0\big[\exp\big(-\lambda L(\tau_y, 0)\big)\big] = (2\lambda y + 1)^{-1} \quad \text{for all} \quad \lambda > 0,$$

from which the desired result follows. For the proof of this relation, let $\lambda > 0$ and for $t \geq 0$, consider

$$\left(\lambda B_t^+ + \frac{1}{2}\right)\exp\big(-\lambda L(t, 0)\big), \quad t \geq 0.$$

We claim this defines a martingale. To see this, set $M_t = \int_0^t 1_{[0,\infty)}(B_s)\,dB_s$, $V_t = L(t, 0)$ and $f(x_1, y_1) = (\lambda x_1 + \frac{\lambda}{2}y_1 + \frac{1}{2})\exp(-\lambda y_1)$. By (7.5), the above expression is equal to $f(M_t, V_t)$. Then by Itô's formula (5.2) and the substitution theorem (Theorem 2.12), it follows that P^0-a.s. for all $t \geq 0$:

$$\left(\lambda B_t^+ + \frac{1}{2}\right)\exp\big(-\lambda L(t, 0)\big) - \frac{1}{2}$$

$$= \lambda \int_0^t \exp\big(-\lambda L(s, 0)\big)1_{[0,\infty)}(B_s)\,dB_s$$

$$+ \frac{\lambda}{2}\int_0^t \exp\big(-\lambda L(s, 0)\big)\,dL(s, 0)$$

$$- \lambda \int_0^t \left(\lambda B_s^+ + \frac{1}{2}\right)\exp\big(-\lambda L(s, 0)\big)\,dL(s, 0).$$

Now, by the last part of Theorem 7.6, almost surely, $L(\,\cdot\,, 0)$ can increase only when B is at zero. It follows that the sum of the last two integrals in the above equation is zero, almost surely. Moreover, the stochastic integral with respect to dB_s there defines an L^2-martingale (cf. Theorem 2.5), hence so does $(\lambda B_t^+ + \frac{1}{2})\exp(-\lambda L(t, 0))$. Then it follows by Doob's stopping theorem that

$$E^0\left[\left(\lambda B_{t\wedge\tau_y}^+ + \frac{1}{2}\right)\exp\big(-\lambda L(t \wedge \tau_y, 0)\big)\right] = \frac{1}{2}.$$

Since $\tau_y < \infty$ and $0 \leq B_{t\wedge\tau_y}^+ \leq y$, P^0-a.s., we may let $t \to \infty$ in the above and conclude by bounded convergence that

$$E^0\left[\left(\lambda y + \frac{1}{2}\right)\exp\big(-\lambda L(\tau_y, 0)\big)\right] = \frac{1}{2}.$$

The desired Laplace transform relation follows from dividing both sides by $(\lambda y + \frac{1}{2})$. ∎

7.4 Proof of Lemma 7.2.

Proof of Lemma 7.2. Let $G(t, x) = \int_0^t 1_{[x,\infty)}(B_s)\, dB_s$. Fix $T > 0$. We first prove that there is a constant C such that for all x and y in $\mathbb{R}$,

$$(7.18) \qquad E\left\{ \sup_{0 \le t \le T} |G(t, x) - G(t, y)|^4 \right\} \le C(x - y)^2.$$

We may suppose $x \le y$. By (6.20),

$$E\left\{ \left(\int_0^t 1_{[x,y)}(B_s)\, dB_s \right)^4 \right\} \le 36 E\left\{ \left(\int_0^t 1_{[x,y)}(B_s)\, ds \right)^2 \right\} \le 36 t^2.$$

Hence, $\{G(t, x) - G(t, y), t \in \mathbb{R}_+\}$ is an L^4-martingale. By Doob's inequality (Theorem 1.4), the left member of (7.18) does not exceed

$$\left(\frac{4}{3} \right)^4 E\left\{ |G(T, x) - G(T, y)|^4 \right\} \le \left(\frac{4}{3} \right)^4 36 E\left\{ \left(\int_0^T 1_{[x,y)}(B_s)\, ds \right)^2 \right\}.$$

The last written expectation may be evaluated as follows. By symmetry and the independence of $B_r - B_0$ and $B_s - B_r$, it is equal to

$$2E\left\{ \int_0^T dr \int_r^T ds\, 1_{[x,y)}(B_s) 1_{[x,y)}(B_r) \right\}$$

$$= 2 \int_0^T dr \int_r^T ds\, (2\pi)^{-1} (r(s - r))^{-1/2}$$

$$\cdot E\left\{ \int_{x-B_0}^{y-B_0} dz \int_{x-B_0}^{y-B_0} dw \exp\left(-\frac{z^2}{2r} - \frac{(w - z)^2}{2(s - r)} \right) \right\}$$

$$\le C_1(x - y)^2,$$

for some constant $C_1 > 0$. This establishes (7.18).

Fix $R > 0$ and let $\mathcal{D}$ denote the set of dyadic numbers of the form $j2^{-n}$ where n and j are integers: $n \geq 0$ and $-R2^n \leq j \leq R2^n$. Put $q = 2^{-\frac{1}{8}}$. We have, by (7.18) and Chebyshev's inequality:

$$P\left\{ \sup_{0 \leq t \leq T} \left| G\left(t, (j+1)2^{-n}\right) - G(t, j2^{-n}) \right| > q^n \right\} \leq q^{-4n} C 2^{-2n}.$$

The sum over all j such that $j2^{-n}$ and $(j+1)2^{-n}$ both belong to $\mathcal{D}$ is bounded by $(2^{n+1}R)C2^{-3n/2} = 2RC2^{-n/2}$. The sum of this over all n is convergent. Thus, by the Borel-Cantelli lemma, there exists Ω_0 with $P(\Omega_0) = 1$ such that for each $\omega \in \Omega_0$, there is $n_0(\omega) \in I\!N$ such that for all $n \geq n_0(\omega)$:

$$(7.19) \qquad \sup_{0 \leq t \leq T} \left| G\left(t, (j+1)2^{-n}\right) - G(t, j2^{-n}) \right| \leq q^n$$

for all j indicated above. From now on we consider ω in Ω_0 only. Let

$$x = \sum_{k=1}^{m} \frac{x_k}{2^k} \quad , \quad y = \sum_{k=1}^{m} \frac{y_k}{2^k}$$

where each x_k or y_k is 0 or 1. We claim that if $|x - y| < 2^{-n}$ where $n \geq n_0(\omega)$, then

$$(7.20) \qquad \sup_{0 \leq t \leq T} |G(t, x) - G(t, y)| \leq 2 \sum_{k=n+1}^{m} q^k.$$

This will be verified by induction on m. If $m \leq n$, then $x = y$ and the sum on the right of (7.20) is zero by the usual convention. Suppose (7.20) is true when m is replaced by $m-1$. Let $x' = \sum_{k=1}^{m-1} x_k 2^{-k}$ and $y' = \sum_{k=1}^{m-1} y_k 2^{-k}$. Then $x - x' = 0$ or 2^{-m}, and $y - y' = 0$ or 2^{-m}. Hence, by (7.19) with $n = m$ we have

$$\sup_{0 \leq t \leq T} |G(t, x) - G(t, x')| \leq q^m, \quad \sup_{0 \leq t \leq T} |G(t, y) - G(t, y')| \leq q^m.$$

By the induction hypothesis,

$$\sup_{0 \leq t \leq T} |G(t, x') - G(t, y')| \leq 2 \sum_{k=n+1}^{m-1} q^k.$$

Adding these three inequalities, we obtain (7.20), and the induction is complete. When $m > n \to \infty$ in (7.20), the sum converges to zero. It follows that for each $(t, x) \in [0, T] \times [-R, R]$,

$$(7.21) \qquad\qquad J(t, x) \equiv \lim_n G(t, x_n)$$

exists for any $\{x_n\} \subset \mathcal{D}$ such that $x_n \to x$. The value of this limit does not depend on the choice of x_n and the convergence is uniform in t. Since $G(t, x_n)$ is continuous in $t \in [0, T]$ for each n, it follows by this uniformity that $J(t, x)$ is also continuous in t for each fixed x. Moreover, as a consequence of (7.20) we have

$$\sup_{0 \le t \le T} |J(t, x) - J(t, y)| \le 2 \sum_{k=n+1}^{\infty} q^k$$

whenever $|x - y| < 2^{-n}$ and $n \ge n_0(\omega)$. Thus if $(t_n, x_n) \to (t, x)$ in $[0, T] \times [-R, R]$, then $J(s, x_n) \to J(s, x)$ uniformly for $s \in [0, T]$, and by the continuity of $J(\cdot, x)$, $J(t_n, x) \to J(t, x)$; consequently $J(t_n, x_n) \to J(t, x)$. This proves J is continuous on $[0, T] \times [-R, R]$. Finally, for any $x \in [-R, R]$ and $t \in [0, T]$, it follows from (7.18) that if $x_n \to x$, then $G(t, x_n)$ converges to $G(t, x)$ in L^4. In conjunction with (7.21) we conclude that $P(G(t, x) = J(t, x)) = 1$, proving (7.6). Since T and R are arbitrary, we can extend the definition of J to $\mathbb{R}_+ \times \mathbb{R}$ on a set of probability one. ∎

7.5 Exercises

1. Prove the claim made in the example at the end of Section 3.3 that $\{\tilde{B}_{L(t)}, \mathcal{F}_{L(t)}, t \ge 0\}$ is a continuous L^2-martingale with quadratic variation process $\{L(t), t \ge 0\}$.

2. Let B and τ_y be defined as in Theorem 7.7. Without assuming the result of that theorem, use (7.5) to verify that

$$E^0[L(\tau_y, 0)] = 2y \quad \text{for all} \quad y \ge 0.$$

It follows from this, by the symmetry and spatial homogeneity of B, that

$$E^x[L(\tau_y, x)] = 2\,|y - x|$$

for all $x, y \in \mathbb{R}$. Note that an attempt to obtain the above relation directly from (7.5) fails when $y < x$. Explain why this is so.

3. Suppose the Brownian motion B satisfies $B(0) = x$ for some $x \in (0, 1)$. Let $\tau = \inf\{t \geq 0 : B_t \notin (0, 1)\}$. Prove that $L(\tau, x)$ is an exponential random variable with mean $2x(1 - x)$.

Hint: Proceed in a similar manner to the proof of Theorem 7.7 using the decompositions of $(B_t - x)^+$ and $(B_t - x)^-$ in place of that for B_t^+.

4. Suppose $B_0 = 0$ a.s. and denote $L(t, 0)$ by L_t. Define $A_t = t + L_t$ for each $t \geq 0$. Observe that $A = \{A_t, t \geq 0\}$ is continuous, adapted and strictly increasing. Define $\tau_t = \inf\{s \geq 0 : A_s > t\}$ for each $t \geq 0$. Observe that $t \to \tau_t$ is continuous, strictly increasing and $A_{\tau_t} = t$. Let $M_t = B_{\tau_t}$. Prove that $\{M_t, \mathcal{F}_{\tau_t}, t \geq 0\}$ is a continuous martingale with quadratic variation process $\{\tau_t, t \geq 0\}$. Show that

$$\int_0^{A_t} 1_{\{0\}}(M_s)ds = L_t \quad \text{for all } t \geq 0.$$

For each $t \geq 0$, verify that A_t is an optional time with respect to the filtration $\{\mathcal{F}_{\tau_s}, s \geq 0\}$. Define

$$Y_t = \int_0^{A_t} 1_{\mathbb{R}\backslash\{0\}}(M_s)dM_s, \quad t \geq 0,$$

and verify that $\{Y_t, \mathcal{F}_t, t \geq 0\}$ is a Brownian motion. (Hint: Use Theorem 6.1.) Loosely speaking, M behaves like a Brownian motion in $\mathbb{R}\backslash\{0\}$, but rather than spending zero Lebesgue time at the origin like Brownian motion, it spends a positive amount of time there. However, M never stays at the origin for an *interval* of time. This process, or sometimes its absolute value, is called a *sticky* Brownian motion.

8

REFLECTED BROWNIAN MOTIONS

8.1 Introduction

In this chapter, the processes $L(\,\cdot\,,0)$ and $\hat{B}(\,\cdot\,,0)$, defined by (7.7) and (7.16), will be denoted respectively by $\hat{B}(\,\cdot\,)$ and $L(\,\cdot\,)$.

By setting $x = 0$ in (7.17) we obtain

$$(8.1) \qquad\qquad |B| = \hat{B} + L.$$

The process $|B|$ is called the reflection of the Brownian motion B at zero. By (8.1), the pair $(|B|, L)$ is almost surely the solution of a certain problem of reflection for $\hat{B}$, discussed in Section 8.2. This yields an alternative representation, directly in terms of $\hat{B}$, for L and hence for $|B|$. The Itô formula will be used in Section 8.3, to make the connection with the analytical theory of this alternative representation. In Section 8.4, we shall give two examples to illustrate how this representation and some of its multi-dimensional analogues arise naturally as diffusion approximations to storage models. A characterization of two-dimensional reflected Brownian motions in a wedge will be discussed in Section 8.5.

8.2 Brownian Motion Reflected at Zero

It follows from (8.1) and Theorem 7.6 that for almost every ω, the pair $(|B|, L)(\omega)$ is a solution of the following problem of reflection for $\hat{B}(\omega)$. The formulation of this problem for real-valued functions without reference to probability is taken from El Karoui and Chaleyat-Maurel [28].

Problem of Reflection. Let C denote the class of continuous functions from $I\!R_+$ to $I\!R$. Given $x \in C$, a pair (z, y) is called a solution of the problem of reflection for x, denoted by $PR(x)$, if $z \in C$, $y \in C$, and the following three conditions are satisfied:

(i) $z = x + y$

(ii) $z \geq 0$

(iii) $y(0) = 0$, y is increasing on $I\!R_+$, and $\int_0^\infty z(t)\, dy(t) = 0$.

The following analytic lemma ensures the existence and uniqueness of a solution to the problem of reflection for any $x \in C$ which satisfies $x(0) \geq 0$.

Lemma 8.1. *Let $x \in C$ with $x(0) \geq 0$. Then $PR(x)$ has a unique solution given by (z, y) where*

$$(8.2) \qquad z = x + y; \quad y(t) = \max_{0 \leq s \leq t} x^-(s) \text{ for each } t \in I\!R_+.$$

Proof. We shall first verify that (z, y), defined by (8.2), is a solution of $PR(x)$. Clearly, y and z are continuous and (i) holds. Condition (ii) is easily verified as follows:

$$z(t) = x(t) + y(t) \geq x(t) + x^-(t) = x^+(t) \geq 0.$$

Since $x(0) \geq 0$, $y(0) = 0$. Obviously y is increasing. To verify that $\int_0^\infty z(t)\, dy(t) = 0$, we must show that t_0 can be a point of increase of y only if $z(t_0) = 0$. Consider the case when y increases to the right of t_0, the proof is similar if the increase is from the left. In this case, for each

$\delta > 0$, $y(t_0) < y(t_0 + \delta)$ and thus, by the definition of y, there must be $t_\delta \in (t_0, t_0 + \delta]$ such that

(8.3) $0 \le y(t_0) < x^-(t_\delta) \le y(t_0 + \delta)$ and hence $x(t_\delta) < 0$.

It follows on letting $\delta \downarrow 0$ in (8.3), by the continuity of x^-, y, and x, that

$$y(t_0) = x^-(t_0) \quad \text{and} \quad x(t_0) \le 0.$$

Hence, $z(t_0) = x(t_0) + y(t_0) = 0$. This completes the proof that (z, y) is a solution of $PR(x)$.

To prove the uniqueness, suppose that $(\hat{z}, \hat{y})$ is another solution of $PR(x)$. Then by condition (i),

(8.4) $z(t) - \hat{z}(t) = y(t) - \hat{y}(t)$

and hence

$$0 \le (y(t) - \hat{y}(t))^2 = 2 \int_0^t (y(s) - \hat{y}(s))\, d(y(s) - \hat{y}(s))$$

$$= 2 \int_0^t (z(s) - \hat{z}(s))\, d(y(s) - \hat{y}(s))$$

$$= -2 \int_0^t z(s)\, d\hat{y}(s) - 2 \int_0^t \hat{z}(s)\, dy(s)$$

$$\le 0$$

where the third equality follows by condition (iii). Thus, $y(t) = \hat{y}(t)$ and hence $z(t) = \hat{z}(t)$ for all t. ∎

Since $(|B|, L)$ is a solution of $PR(\hat{B})$, a.s., and $\hat{B}(0) = |B_0| \ge 0$, it follows by the uniqueness part of Lemma 8.1 that almost surely:

(8.5) $L_t = \max_{0 \le s \le t} \hat{B}_s^-$ for all t in $\mathbb{R}_+$.

Definition. Two processes X and Y are said to be equivalent in law if they have the same finite dimensional distributions, i.e., if

$$P(X_{t_1} \leq x_1, X_{t_2} \leq x_2, \ldots, X_{t_n} \leq x_n)$$
$$= P(Y_{t_1} \leq x_1, Y_{t_2} \leq x_2, \ldots, Y_{t_n} \leq x_n)$$

for all $0 \leq t_1 < t_2 < \ldots < t_n < \infty$; $x_1, x_2, \ldots, x_n$ in $I\!R$; and $n \in I\!N$.

It follows from the proof of Theorem 6.1 that $\hat{B} - \hat{B}_0$ is independent of $\mathcal{F}_0 = \sigma(B_0)^\sim$. Moreover, $\hat{B} - \hat{B}_0$ is a Brownian motion starting at zero and is therefore equivalent in law to $B - B_0$ and by symmetry to $-(B - B_0)$, where the latter two processes are also independent of $\mathcal{F}_0$. These properties, together with the fact that $|\hat{B}_0| = |B_0|$, imply that $|\hat{B}|$ is equivalent in law to $|B|$.

It is important to realize that equivalence in law of two processes is a weaker property than equality of sample paths for each ω. For example, $|\hat{B}|$ is equivalent in law to $|B|$, but $|\hat{B}| \neq |B|$. For if $|\hat{B}|$ were equal to $|B|$, then (8.1) would read $|\hat{B}| = \hat{B} + L$ where $|\hat{B}|$ and $\hat{B}$ are simultaneously zero for a sequence of times tending to ∞, and this would imply the false conclusion $L \equiv 0$.

Definition. For each t let

$$(8.6) \qquad Z_t = \hat{B}_t + \max_{0 \leq s \leq t} \hat{B}_s^-.$$

By (8.1) and (8.5), $Z = |B|$ almost surely and hence Z is equivalent in law to $|\hat{B}|$. Moreover, by (8.5) and (7.1), we have almost surely:

$$(8.7) \qquad \begin{aligned} \max_{0 \leq s \leq t} \hat{B}_s^- &= L_t = \lim_{\varepsilon \downarrow 0} \frac{1}{2\varepsilon} \lambda \{0 \leq s \leq t : |B_s| \leq \varepsilon\}. \\ &= \lim_{\varepsilon \downarrow 0} \frac{1}{2\varepsilon} \lambda \{0 \leq s \leq t : Z_s \leq \varepsilon\}. \end{aligned}$$

It is therefore natural that L is also referred to as the local time of Z at zero. At the end of this chapter, (8.7) will be derived by the alternative means of a direct calculation.

In the above we have seen two ways of representing the process obtained by reflecting the Brownian motion B at zero. The first is $|B|$ and the second is Z. The second is a useful representation, both for analytical reasons and from the point of view of applications. We illustrate below how the connection with the analytical theory of Z can be made via the Itô formula. This is followed by a discussion of applications in storage theory in which Z and more general reflected processes arise as diffusion approximations.

8.3 Analytical Theory of Z via the Itô Formula

We can easily apply the Itô formula to Z because it is the sum of the continuous local martingale $\hat{B}$ and the continuous increasing process L. For example, we have the following.

Theorem 8.2. *Let $\phi \in C^2(\mathbb{R})$ and $\alpha \geq 0$. Then a.s. for all t we have*

$$
\begin{aligned}
e^{-\alpha t}\phi(Z_t) - \phi(Z_0) &= \int_0^t e^{-\alpha s}\phi'(Z_s)\,d\hat{B}_s + \int_0^t e^{-\alpha s}\phi'(Z_s)\,dL_s \\
&\quad - \alpha \int_0^t \phi(Z_s)e^{-\alpha s}\,ds + \frac{1}{2}\int_0^t e^{-\alpha s}\phi''(Z_s)\,ds.
\end{aligned}
$$

(8.8)

Proof. This follows by applying Theorem 5.10 with $M_t = \hat{B}_t$, $V_t^1 = L_t$, $V_t^2 = e^{-\alpha t}$, domain $D = \mathbb{R}^3$, and function f defined by $f(x, y_1, y_2) = y_2\phi(x + y_1)$ for all (x, y_1, y_2) in $\mathbb{R}^3$. ∎

Corollary 8.3. *Suppose $\phi \in C^2(\mathbb{R})$, $h \in C(\mathbb{R}_+)$, and $\alpha \geq 0$, are such that*

$$
(8.9) \qquad \alpha\phi - \frac{1}{2}\phi'' = h \text{ on } \mathbb{R}_+ \quad \text{and} \quad \phi'(0) = 0.
$$

Then $M = \{e^{-\alpha t}\phi(Z_t) - \phi(Z_0) + \int_0^t e^{-\alpha s}h(Z_s)\,ds, t \in \mathbb{R}_+\}$ is a continuous local martingale. Furthermore, if ϕ and h are bounded on $\mathbb{R}_+$, $\alpha > 0$, and $B_0 = x \geq 0$, then

$$
(8.10) \qquad \phi(x) = E^x\left(\int_0^\infty e^{-\alpha s}h(Z_s)\,ds\right).
$$

Proof. Since L can only increase when Z is zero and $\phi'(0) = 0$, the integral with respect to dL_s in (8.8) is zero for all t. Then it follows from (8.8) and (8.9), since $Z \geq 0$, that

$$(8.11) \qquad e^{-\alpha t}\phi(Z_t) - \phi(Z_0) + \int_0^t e^{-\alpha s} h(Z_s)\, ds = \int_0^t e^{-\alpha s}\phi'(Z_s)\, d\hat{B}_s$$

where the stochastic integral on the right defines a continuous local martingale. This proves the first assertion.

Suppose that the hypotheses preceding (8.10) are satisfied. Since ϕ and h are bounded, then M_t, given by the left side of (8.11), is uniformly bounded for all t in any bounded interval. It follows by Proposition 1.8 that M is a martingale. Hence, $E^x(M_t) = E^x(M_0) = 0$. Let $t \to \infty$, then (8.10) follows by bounded convergence. ∎

Remark. Since $Z_0 = |B_0| = x$, the expectation E^x in (8.10) can be interpreted as the expectation given $Z_0 = x$.

By using (8.6), it can be verified that for any finite-valued optional time τ, and $t \geq 0$,

$$(8.12) \qquad Z_{\tau+t} = Z_\tau + \hat{B}_{\tau+t} - \hat{B}_\tau + \max_{0 \leq s \leq t}(Z_\tau + \hat{B}_{\tau+s} - \hat{B}_\tau)^-.$$

The proof of Theorem 6.1 can be modified to show that $\{\hat{B}_{\tau+t} - \hat{B}_\tau, t \in \mathbb{R}_+\}$ is a Brownian motion independent of $\mathcal{F}_\tau$. Then it follows from (8.12) that Z is a strong Markov process. Let $b\mathcal{B}$ denote the class of all bounded $\mathcal{B}$-measurable functions $g : \mathbb{R}_+ \to \mathbb{R}$. In the theory of Markov processes, the operator $R_\alpha : b\mathcal{B} \to b\mathcal{B}$ defined for $\alpha > 0$ by

$$(R_\alpha g)(x) = E^x \left(\int_0^\infty e^{-\alpha s} g(Z_s)\, ds \right)$$

for all $x \geq 0$ and $g \in b\mathcal{B}$, is called the α-*resolvent* of Z. It follows from Corollary 8.3 that given a bounded function $h \in C(\mathbb{R}_+)$ and $\alpha > 0$, if there is a bounded function $\phi \in C^2(\mathbb{R})$ satisfying (8.9), then $\phi = R_\alpha h$. For instance, if h vanishes outside a bounded interval, then by the theory of ordinary differential equations such a ϕ exists and by the representation (8.10) it is unique. Thus, in a sense (which can be made precise), R_α is the inverse of the differential operator $\alpha - \frac{1}{2}d^2/dx^2$.

We have emphasized the decomposition $Z = \hat{B} + L$ and the use of the Itô formula in making the connection with the analytical theory of Z. This approach generalizes to more complex reflected Brownian motions, such as those with two reflecting barriers (one at $x = 0$, the other at $x = b > 0$) or those in multi-dimensions (see Example 2 below). There is no simple alternative representation such as $|B|$ for these more general processes.

8.4 Approximations in Storage Theory

The process Z arises naturally in applications as a diffusion approximation in storage theory. This will be illustrated in Example 1 below with a model of a single storage facility, where Z features in the limit theory of the contents process.

More generally, Brownian motions with two reflecting barriers and multi-dimensional reflected Brownian motions also arise in the approximation of storage systems. Such processes behave like Brownian motions inside convex polyhedral domains in $\mathbb{R}^d, d \geq 1$, and are reflected from the boundaries of these domains, where the direction of reflection is constant on each boundary hyperplane. In Example 2, we shall discuss a model of two storage facilities in series, where the two-dimensional contents process can be approximated by a reflected Brownian motion on the positive quadrant $\mathbb{R}^2_+$. For this process, the direction of reflection is normal on one boundary half-line and oblique on the other.

Example 1. *Model of a single storage facility.*

Consider a single storage facility, such as a buffer or dam, with infinite capacity. Because of its intuitive appeal, we shall imagine the facility to be a dam for storing water. Suppose the dam is initially empty, inflow to the dam is random, and at times $n \in \mathbb{N}$ there are demands for release of water from the dam. A demand is fulfilled only up to the contents of the dam at the time; thus the actual amount of water released at time n is the smaller of the demand and the contents of the dam at that time. The following probabilistic setup will be interpreted as a model of this storage facility.

Let $\{A_n, n \in I\!N\}$ and $\{R_n, n \in I\!N\}$ be two independent sequences of independent identically distributed (i.i.d.) positive random variables with finite, strictly positive, means and variances, such that $E(A_n) = E(R_n)$ for all n. For each n let $U_n = A_n - R_n$. Then $E(U_n) = 0$ and $\sigma^2 \equiv E(U_n^2) > 0$. Let $\{W_n, n \in I\!N_0\}$ be the sequence of r.v.'s defined inductively by

(8.13) $W_0 = 0$, $W_n = (W_{n-1} + U_n)^+$ for each $n \in I\!N$.

If A_n represents the amount of water which has flowed into the dam during the time interval $(n-1, n]$ and R_n represents the amount of water demanded for release at time n, then U_n can be interpreted as the potential change in the amount of water in the dam during the time interval $(n-1, n]$. We say "potential" because this change may not be fully realized, since W_n, which can be interpreted as the amount of water in the dam after the release at time n, must be non-negative.

Let $\{X_n, n \in I\!N_0\}$ be the sequence of r.v.'s defined by

$$X_0 = 0 \ , \ X_n = \sum_{i=1}^{n} U_i \text{ for each } n \in I\!N.$$

The r.v. X_n can be interpreted as the cumulative potential change in the amount of water in the dam up to time n. The following expression for W_n in terms of $\{X_i : 0 \leq i \leq n\}$ can be obtained by induction on n using (8.13):

(8.14) $W_n = X_n + \max_{0 \leq i \leq n} (-X_i)$ for each $n \in I\!N_0$.

For each $n \in I\!N$, let $W^n = \{W_t^n, t \in I\!R_+\}$ be the process defined by

$$W_t^n = \left(\sigma \sqrt{n}\right)^{-1} W_{[nt]} \text{ for each } t \geq 0.$$

We shall continue to use the symbols $\hat{B}$ and Z as in the preceding part of this chapter. Recall that $\hat{B}$ is a Brownian motion in $I\!R$ and Z is defined by (8.6). Moreover, we shall assume for the duration of this example that $\hat{B}_0 = 0$. The focal point is the following limit theorem.

Theorem 8.4. *For each t, W_t^n converges in distribution to Z_t as $n \to \infty$.*

Sketch of proof. The main omission in the following is of details relating to weak convergence in the function space $D[0, m]$ of all functions defined on $[0, m]$, $m \in I\!N$, which are right continuous on $[0, m)$ and have finite left limits on $(0, m]$. For the missing details we refer the reader to Billingsley [3]. Actually, Billingsley only refers to $D[0, 1]$ but a simple dilation of the time scale yields the same theory for $D[0, m]$. We use the abbreviation "in dist." to mean "in distribution".

By the central limit theorem, for each $t > 0$, as $n \to \infty$,

$$\sqrt{[nt]/n} \left(\sigma \sqrt{[nt]} \right)^{-1} \sum_{i=1}^{[nt]} U_i \to \hat{B}_t \quad \text{in dist.}$$

It follows that for each $t \geq 0$,

$$X_t^n \equiv \left(\sigma \sqrt{n} \right)^{-1} X_{[nt]} \to \hat{B}_t \quad \text{in dist.}$$

In fact for each $m \in I\!N$, $\{X_t^n, 0 \leq t \leq m\}$ converges weakly in $D[0, m]$ to $\{\hat{B}_t, 0 \leq t \leq m\}$. Then it follows by the continuous mapping theorem [3; Corollary 1, p. 31] that

$$\{W_t^n = X_t^n + \max_{0 \leq s \leq t} (-X_s^n), 0 \leq t \leq m\}$$

converges in the same weak sense to

$$\{\hat{B}_t + \max_{0 \leq s \leq t} (-\hat{B}_s), 0 \leq t \leq m\}.$$

Since $\hat{B}_0 = 0$, we have

$$\max_{0 \leq s \leq t} (-\hat{B}_s) = \max_{0 \leq s \leq t} \hat{B}_s^-.$$

Weak convergence on $D[0, m]$ for each $m \in I\!N$ implies convergence in distribution for each time t, hence

$$W_t^n \to Z_t = \hat{B}_t + \max_{0 \leq s \leq t} \hat{B}_s^- \quad \text{in dist.}$$

for each t. ∎

It follows from this theorem and the definition of W_t^n that Z can be used to approximate the long run behavior of W_n.

The process Z can also be employed as an approximation to the waiting time sequence of a single-server queue in heavy traffic. A single-server queue is a service facility with one server at which customers arrive individually and queue up for service which is on a first-come-first-served basis. Suppose, for each $n \in I\!N$, that the r.v. A_n of the preceding example represents the service time of the nth customer to arrive at a single-server queue, and that R_n represents the time between the arrival of the nth and the $(n + 1)^{\text{st}}$ customer. Then W_n can be interpreted as the waiting-time (excluding service time) of the $(n + 1)^{\text{st}}$ customer, assuming that the first customer did not have to wait. With these interpretations, the model is referred to as a $GI/G/1$ queue with traffic intensity $\rho \equiv E(A_n)/E(R_n) = 1$. Thus by Theorem 8.4, Z can be used to approximate the long run behavior of the waiting time sequence of a $GI/G/1$ queue with traffic intensity one.

Example 2. *Model of two storage facilities in series.*

Consider two storage facilities in series, each having infinite capacity. We shall again refer to these facilities as dams. Suppose that both dams are initially empty and that there is random inflow to the first dam, while at discrete times $n \in I\!N$, there are demands for transfer of a fixed constant amount of water from the first dam to the second dam and for release of water from the second dam. At time n, water is transferred from the first to the second dam and then water is released from the second dam, to fulfill the transfer and release demands as far as the contents of the dams will allow. The flow of water is illustrated in Figure 8.1.

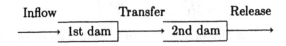

Figure 8.1.

The following probabilistic setup will be interpreted as a model of this system.

Let $\{A_n, n \in I\!N\}$ and $\{R_n, n \in I\!N\}$ be two independent sequences of i.i.d. positive random variables with finite, strictly positive, means and variances, such that $E(A_n) = E(R_n) = c$ for all n and some constant $c > 0$. Let $\{T_n, n \in I\!N\}$ denote the deterministic process defined by $T_n = c$ for all n. Let

$$U_n^1 = A_n - T_n \ , \ U_n^2 = T_n - R_n.$$

Then $E(U_n^1) = 0$, $E(U_n^2) = 0$, $\sigma_1^2 \equiv E\{(U_n^1)^2\} > 0$, and $\sigma_2^2 \equiv E\{(U_n^2)^2\} > 0$. Let $\{W_n^1, n \in I\!N_0\}$ and $\{W_n^2, n \in I\!N_0\}$ be the sequences of r.v.'s defined inductively by

(8.15) $\qquad W_0^1 = 0 \ , \ W_n^1 = (W_{n-1}^1 + U_n^1)^+$ for $n \geq 1,$

(8.16) $\qquad \begin{cases} W_0^2 = 0, \\ W_n^2 = (W_{n-1}^2 + \min(T_n, W_{n-1}^1 + A_n) - R_n)^+ \\ \qquad\qquad\qquad\qquad\qquad \text{for each } n \geq 1. \end{cases}$

If A_n represents the amount of water which has flowed into the first dam during the time interval $(n-1, n]$, and T_n represents the fixed constant amount demanded for transfer at time n; then W_n^1 can be interpreted as the amount of water in the first dam after the transfer of water at time n, and $\min(T_n, W_{n-1}^1 + A_n)$ represents the actual amount transferred at that time. If in addition R_n represents the amount of water demanded for release from the second dam at time n, then W_n^2 can be interpreted as the amount of water in the second dam after the release of water at that time.

For each $j \in \{1, 2\}$, let $\{X_n^j, n \in I\!N_0\}$ be the sequence defined by

(8.17) $\qquad X_0^j = 0 \ , \ X_n^j = \sum_{i=1}^{n} U_i^j$ for each $n \in I\!N.$

Let $\{Y_n^1, n \in I\!N_0\}$ and $\{Y_n^2, n \in I\!N_0\}$ be defined by

(8.18) $\qquad Y_n^1 = \max_{0 \leq i \leq n}(-X_i^1) \ , \ Y_n^2 = \max_{0 \leq i \leq n}(-X_i^2 + Y_i^1).$

The following lemma gives useful alternative expressions for W_n^1 and W_n^2.

Lemma 8.5. *For each* $n \in \mathbb{N}_0$,

$$(8.19) \qquad W_n^1 = X_n^1 + Y_n^1 \quad \text{and}$$

$$(8.20) \qquad W_n^2 = X_n^2 - Y_n^1 + Y_n^2.$$

Proof. Since W_n^1 is defined in the same recursive manner as W_n in Example 1, (8.19) follows from (8.14). We shall use (8.19) and induction to prove (8.20).

When $n = 0$, both sides of (8.20) are zero. Suppose $n > 0$ and (8.20) holds with $n - 1$ in place of n. Then by (8.16) :

$$W_n^2 = \left(W_{n-1}^2 + \min(T_n, W_{n-1}^1 + A_n) - R_n\right)^+$$
$$= \left(W_{n-1}^2 + T_n - R_n + \min(0, W_{n-1}^1 + A_n - T_n)\right)^+.$$

By the induction assumption, the definitions of U_n^1 and U_n^2, and (8.19), the last line above equals

$$\left(X_{n-1}^2 - Y_{n-1}^1 + Y_{n-1}^2 + U_n^2 + \min(0, X_{n-1}^1 + Y_{n-1}^1 + U_n^1)\right)^+$$
$$= \left(X_n^2 + Y_{n-1}^2 + \min(-Y_{n-1}^1, X_n^1)\right)^+ \quad \text{by (8.17)}$$
$$= \left(X_n^2 + Y_{n-1}^2 - \max(Y_{n-1}^1, -X_n^1)\right)^+$$
$$= \left(X_n^2 - Y_n^1 + Y_{n-1}^2\right)^+ \quad \text{by (8.18)}$$
$$= \max\left(X_n^2 - Y_n^1 + Y_{n-1}^2, 0\right)$$
$$= X_n^2 - Y_n^1 + \max\left(Y_{n-1}^2, -(X_n^2 - Y_n^1)\right)$$
$$= X_n^2 - Y_n^1 + Y_n^2.$$

Thus (8.20) holds for n. This completes the induction proof. ∎

To state the analogue of Theorem 8.4, we need the notion of convergence in distribution of pairs of r.v.'s.

Definition. A sequence $\{(V_n^1, V_n^2)\}$ of pairs of r.v.'s converges *in distribution* (in dist.) to a pair of r.v.'s (V^1, V^2) if

$$\lim_{n \to \infty} P\left(V_n^1 \leq x_1, V_n^2 \leq x_2\right) = P\left(V^1 \leq x_1, V^2 \leq x_2\right)$$

for each $(x_1, x_2) \in \mathbb{R}^2$ which is a point of continuity of the function given by the right side of this equality.

For each $n \in \mathbb{N}$ let $\mathbf{W}^n$ be the two-dimensional process defined by

$$\mathbf{W}_t^n = \frac{1}{\sqrt{n}}\left(\frac{1}{\sigma_1}W_{[nt]}^1, \frac{1}{\sigma_2}W_{[nt]}^2\right) \quad \text{for all } t \geq 0.$$

For simplicity, we assume $\sigma_1 = \sigma_2$ for the remainder of this section. (Analogous results to those below hold for $\sigma_1 \neq \sigma_2$. They have $-(\sigma_1/\sigma_2)L^1$ in place of $-L^1$ in the expressions associated with Z^2.)

Let $\mathbf{B}$ denote a Brownian motion in $\mathbb{R}^2$ with initial value $\mathbf{B}_0 = (0,0)$. We define below the two-dimensional processes $\mathbf{L}$ and $\mathbf{Z}$ by continuous mappings of the paths of $\mathbf{B}$. These mappings are the analogues for continuous time of (8.18)–(8.20).

Definitions. Let $\mathbf{L} = (L^1, L^2)$ and $\mathbf{Z} = (Z^1, Z^2)$ be the two-dimensional processes defined by:

(8.21) $$L_t^1 = \max_{0 \leq s \leq t}(B_s^1)^-, \quad L_t^2 = \max_{0 \leq s \leq t}(B_s^2 - L_s^1)^-$$

(8.22) $$Z_t^1 = B_t^1 + L_t^1, \quad Z_t^2 = B_t^2 - L_t^1 + L_t^2$$

for each $t \geq 0$. Note that

$$(B_s^1)^- = -B_s^1 \quad \text{and} \quad (B_s^2 - L_s^1)^- = -B_s^2 + L_s^1,$$

since $\mathbf{B}_0 = (0,0)$.

Remark. Since the process $\mathbf{L} = (L^1, L^2)$ only appears in this example, there should be no confusion with the spaces of integrable functions L^1 and L^2.

Theorem 8.6. *For each t, as $n \to \infty$:*

(8.23) $\mathbf{W}_t^n \to \mathbf{Z}_t$ *in dist.*

Sketch of proof. The method of proof is similar to that for Theorem 8.4. Since the sequences $\{X_n^1, n \in \mathbb{N}\}$ and $\{X_n^2, n \in \mathbb{N}\}$ are independent, it follows by applying the usual central limit theorem to each component of $\mathbf{X}_t^n$ separately that for each t:

$$\mathbf{X}_t^n \equiv \frac{1}{\sqrt{n}} \left(\frac{1}{\sigma_1} X_{[nt]}^1, \frac{1}{\sigma_2} X_{[nt]}^2 \right) \to \mathbf{B}_t = (B_t^1, B_t^2) \quad \text{in dist.}$$

In fact it can be shown that $\{\mathbf{X}_t^n, \ 0 \le t \le m\}$ converges weakly on $D[0, m] \times D[0, m]$ to $\{\mathbf{B}_t, \ 0 \le t \le m\}$, and then by the continuous mapping theorem and (8.18)–(8.22), $\{\mathbf{W}_t^n, \ 0 \le t \le m\}$ converges in the same weak sense to $\{\mathbf{Z}_t, \ 0 \le t \le m\}$ for each $m \in \mathbb{N}$. The result (8.23) follows from this. ∎

If the definition of $\mathbf{B}_0$ is extended to allow the initial value of $\mathbf{B}_0$ to be $x \in \mathbb{R}_+^2$, not just 0, then the definitions (8.21)–(8.22) of $\mathbf{L}$ and $\mathbf{Z}$ still make sense and imply $\mathbf{Z}_0 = \mathbf{B}_0$. In the following discussion, we consider $\mathbf{B}, \mathbf{L},$ and $\mathbf{Z}$, to be defined in this extended sense, with $\mathbf{B}_0 = x$ for some $x \in \mathbb{R}_+^2$. Then by comparing (8.21)–(8.22) with (8.2) we see that for each ω, $(Z^1, L^1)(\omega)$ is the solution of the problem of reflection for $B^1(\omega)$, and $(Z^2, L^2)(\omega)$ is the solution of the problem of reflection for $(B^2 - L^1)(\omega)$. It follows from the properties of these solutions that

(i) $\begin{pmatrix} Z_t^1 \\ Z_t^2 \end{pmatrix} = \begin{pmatrix} B_t^1 \\ B_t^2 \end{pmatrix} + \begin{pmatrix} 1 & 0 \\ -1 & 1 \end{pmatrix} \begin{pmatrix} L_t^1 \\ L_t^2 \end{pmatrix}$ for each t,

(ii) $Z_t^j \ge 0$ for $j = 1, 2$, and each t,

(iii) $L_0^j = 0$, L^j is increasing, $\int_0^\infty Z_t^j \, dL_t^j = 0$ for $j = 1, 2$.

In fact, because the equation $Z^1 = B^1 + L^1$ only involves the first components of the processes $\mathbf{Z}, \mathbf{B},$ and $\mathbf{L}$, one can use the uniqueness of the solution of the problem of reflection for $B^1(\omega)$ and then the uniqueness of the solution of the problem of reflection for $(B^2 - L^1)(\omega)$ to show that given $\mathbf{B}$, the pair $(\mathbf{Z}, \mathbf{L})$ is the unique pair of continuous two-dimensional processes satisfying (i)→(iii). The matrix $\begin{pmatrix} 1 & 0 \\ -1 & 1 \end{pmatrix}$ is called the reflection matrix.

Properties (i)→(iii) yield a geometric interpretation of the sample path behavior of $\mathbf{Z}$. By (ii), $\mathbf{Z}$ is confined to the positive quadrant $I\!R_+^2$. Property (iii) implies that L^1 can increase only when Z^1 is at zero and L^2 can increase only when Z^2 is at zero. Hence by (i), $\mathbf{Z}$ behaves like a Brownian motion away from the boundary half-lines of $I\!R_+^2$ and when L^1 increases, $\mathbf{Z}$ is reflected (or perhaps more appropriately deflected) from the boundary $z_1 = 0$ in the direction given by the first column of the reflection matrix. Similarly, when L^2 increases, $\mathbf{Z}$ is reflected from the boundary $z_2 = 0$ in the direction given by the second column of this matrix, which happens to be the direction normal to $z_2 = 0$. Thus the directions of reflection of $\mathbf{Z}$ on each boundary half-line are as shown in Figure 8.2.

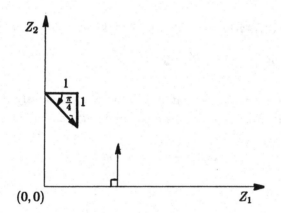

Figure 8.2.

Just as for the one-dimensional process Z (see Theorem 8.2), we can easily apply the Itô formula to $\mathbf{Z}$ to make the connection with its analytical theory. For example we have the following two results.

Theorem 8.7. *Let ψ be a twice continuously differentiable function defined on $I\!R^2$ and $\alpha \geq 0$. Then a.s. for all t we have:*

$$
e^{-\alpha t}\psi(\mathbf{Z}_t) - \psi(\mathbf{Z}_0) = \sum_{j=1}^{2} \int_0^t e^{-\alpha s} \frac{\partial \psi}{\partial z_j}(\mathbf{Z}_s)\, dB_s^j
$$

$$
+ \int_0^t e^{-\alpha s} \left\{ \frac{\partial \psi}{\partial z_1}(\mathbf{Z}_s) - \frac{\partial \psi}{\partial z_2}(\mathbf{Z}_s) \right\} dL_s^1
$$

(8.24)

$$
+ \int_0^t e^{-\alpha s} \frac{\partial \psi}{\partial z_2}(\mathbf{Z}_s)\, dL_s^2
$$

$$
- \alpha \int_0^t \psi(\mathbf{Z}_s)e^{-\alpha s}\, ds + \frac{1}{2} \int_0^t \Delta\psi(\mathbf{Z}_s)\, ds.
$$

Proof. This follows by applying Theorem 5.10 with $M_t^1 = B_t^1$, $M_t^2 = B_t^2$, $V_t^1 = L_t^1$, $V_t^2 = L_t^2$, $V_t^3 = e^{-\alpha t}$, domain $D = I\!R^5$, and function f defined by

$$
f(x_1, x_2, y_1, y_2, y_3) = y_3\psi(x_1 + y_1, x_2 - y_1 + y_2),
$$

for all $(x_1, x_2, y_1, y_2, y_3)$ in $I\!R^5$. ∎

Corollary 8.8. *Suppose that ψ is a twice continuously differentiable function on $I\!R^2$, h is a continuous function on $I\!R_+^2$, and $\alpha \geq 0$, such that*

(8.25) $$\alpha\psi - \frac{1}{2}\Delta\psi = h \quad \text{on } I\!R_+^2$$

and

(8.26)
$$
\begin{cases}
\left(\dfrac{\partial \psi}{\partial z_1} - \dfrac{\partial \psi}{\partial z_2} \right)(0, z_2) = 0 & \text{for } z_2 \geq 0 \\[2mm]
\dfrac{\partial \psi}{\partial z_2}(z_1, 0) = 0 & \text{for } z_1 \geq 0.
\end{cases}
$$

Then $\{e^{-\alpha t}\psi(\mathbf{Z}_t) - \psi(\mathbf{Z}_0) + \int_0^t e^{-\alpha s}h(\mathbf{Z}_s)\, ds, t \in I\!R_+\}$ is a continuous local martingale. Furthermore, if ψ and h are bounded on $I\!R_+^2$, $\alpha > 0$, and

$\mathbf{B}_0 = x \geq 0$, *then*

$$\psi(x) = E^x \left(\int_0^\infty e^{-\alpha s} h(\mathbf{Z}_s)\, ds \right).$$

Proof. Since L^j increases only when Z^j is zero, for $j = 1, 2$, it follows from (8.26) that the second and third integrals in (8.24) are zero. The remainder of the proof is similar to that of Corollary 8.3. ∎

We can rewrite (8.26) in the vector form

$$\begin{cases} \nabla \psi \cdot \left({}^{\ 1}_{-1} \right) = 0 & \text{on } \{(z_1, z_2) : z_1 = 0, z_2 \geq 0\} \\ \nabla \psi \cdot \left({}^0_1 \right) = 0 & \text{on } \{(z_1, z_2) : z_1 \geq 0, z_2 = 0\} \end{cases}$$

where $\nabla \psi = \left(\frac{\partial \psi}{\partial z_1}, \frac{\partial \psi}{\partial z_2} \right)$. This makes it clear that on each boundary half-line of $\mathbb{R}^2_+$, the required boundary condition is that the directional derivative of ψ be zero in the direction of reflection for that half-line (see Figure 8.2).

Further Reading. Reflected Brownian motions in d-dimensional polyhedral domains $(d \geq 2)$, with some restrictions on the directions of reflection, arise as approximations to the waiting time sequences and queue length processes for networks of queues in heavy traffic. For example, details of the approximation for open queueing networks, leading to reflected Brownian motions in the positive orthant, can be found in Reiman [63] (see [43] for an overview), and the associated analytical theory is discussed in Harrison and Reiman [42].

For such reflected Brownian motions there is a major difficulty with the approach to the analytical theory used in Corollary 8.8 and [42]. Namely, it is a non-trivial problem in partial differential equations to prove that equations such as (8.25)–(8.26) have a sufficiently large family of solutions, with the required regularity, to characterize the associated process. The difficulty arises because there is a discontinuity in the directions of reflection across the unsmooth parts of the boundary of the state space. In [75], Varadhan and Williams have given an alternative characterization of two-dimensional reflected Brownian motions in a wedge where the directions of reflection are constant along each side of the wedge. This characterization, described in the next section, is a combination of probabilistic and analytic aspects of the processes. It facilitates their study, and it is valid even when no simple path-by-path construction or natural diffusion approximation is available.

8.5 Reflected Brownian Motions in a Wedge

Consider the problem of existence and uniqueness of a strong Markov process $\mathbf{Z}$ that has continuous sample paths and the following additional properties.

(1)   The state space S is an infinite two-dimensional wedge, and $\mathbf{Z}$ behaves in the interior like Brownian motion.

(2)   The process $\mathbf{Z}$ reflects instantaneously at the boundary of the wedge, the direction of reflection being constant along each side.

(3)   The amount of time that $\mathbf{Z}$ spends at the corner of the wedge is zero (in the sense of Lebesgue measure).

Without loss of generality, we may suppose that the corner of the wedge is at the origin and one side is along the positive z_1-axis, and the angle ξ of the wedge is in $(0, 2\pi)$. The two sides of the wedge will be denoted by ∂S_1 and ∂S_2, and the directions of reflection on these sides will be denoted by constant vectors v_1 and v_2, having positive components in the directions of the inward normals to their respective sides. Associated with these directions of reflection are angles of reflection θ_1 and θ_2, defined as the angles between the inward normals to the sides and the associated directions of reflection. Each of these angles takes its value in $(-\frac{\pi}{2}, \frac{\pi}{2})$. The sign convention for each angle is that it is positive if the associated direction of reflection points towards the corner. A typical acute wedge is drawn in Figure 8.3. The θ_1 and θ_2 shown there are both positive.

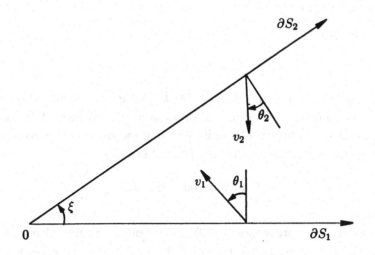

Figure 8.3.

Example. The process **Z** discussed in Example 2 of the preceding section is a continuous strong Markov process satisfying (1)–(3) with

$$\xi = \frac{\pi}{2}, \quad v_1 = \begin{pmatrix} 0 \\ 1 \end{pmatrix}, \quad v_2 = \begin{pmatrix} 1 \\ -1 \end{pmatrix}, \quad \theta_1 = 0, \quad \text{and } \theta_2 = \frac{\pi}{4}.$$

The above formulation is a little heuristic in that one can ask what it means for the process **Z** to behave like Brownian motion in the interior of the wedge, or to reflect instantaneously at the boundary in a given direction. In [75], the question of existence and uniqueness is recast more precisely as the following submartingale problem, in the style used by Stroock and Varadhan [72] for diffusions on smooth domains with smooth boundary conditions. Here C_S denotes the space of continuous functions $z : [0, \infty) \to S$, and for each $t \geq 0$, $\mathcal{M}_t = \sigma\{z(s) : 0 \leq s \leq t\}$ is the smallest σ-field of subsets of C_S such that each of the coordinate maps $z \to z(s)$ is $\mathcal{M}_t$-measurable for $0 \leq s \leq t$. We let $\mathcal{M} \equiv \bigvee_{t \geq 0} \mathcal{M}_t$.

Submartingale Problem. A family of probability measures $\{P^z, z \in S\}$ on C_S is called a solution of the submartingale problem if for each $z \in S$ the following three properties hold.

(i) $P^z\big(z(0) = z\big) = 1.$

(ii) $E^{P^z}\left[\int_0^\infty 1_{\{0\}}\big(z(s)\big)\, ds\right] = 0.$

(iii) $\left\{f\big(z(t)\big) - \frac{1}{2}\int_0^t \Delta f\big(z(s)\big)\, ds, \mathcal{M}_t, t \geq 0\right\}$ is a P^z-submartingale for each function f which is constant in a neighborhood of the origin, and which together with its first and second partial derivatives is continuous and bounded on $I\!R^2$, and satisfies

$$\nabla f \cdot v_j \geq 0 \quad \text{on} \quad \partial S_j \quad \text{for} \quad j = 1, 2.$$

The heuristic properties (1)–(3) are evident in this formulation. Property (1) is mirrored in the choice of the probability space C_S and the presence in (iii) of $\frac{1}{2}\Delta$, the infinitesimal generator of Brownian motion. Property (2) is represented by the directional derivative condition on f. Property (3) follows from condition (ii). Moreover, if the submartingale problem has a unique solution, then the strong Markov property follows.

Example. Consider the process $\mathbf{Z}$ in Example 2 of the preceding section and for each $z \in S$, let Q^z denote a probability measure under which $\mathbf{B}$ is a Brownian motion starting from z. It can be verified, with the aid of the Itô formula, that the family of probability measures $\{P_{\mathbf{Z}}^z, z \in S\}$, defined on C_S by

$$P_{\mathbf{Z}}^z\big(z(\,\cdot\,) \in A\big) = Q^z\big(\mathbf{Z}(\,\cdot\,) \in A\big)$$

for each $A \in \mathcal{M}$, is a solution of the submartingale problem.

In general, we have the following result which is proved in [75]. Define

$$\alpha \equiv (\theta_1 + \theta_2)/\xi.$$

Theorem 8.9. *If $\alpha < 2$, there is a unique solution of the submartingale problem. If $\alpha \geq 2$, there is no solution; however, there is a unique solution* *provided condition (ii) is omitted, which corresponds to the process with absorption at the corner.*

We can only have $\alpha \geq 2$ on acute wedges, i.e., for which $0 < \xi < \frac{\pi}{2}$. However, some feeling for what $\alpha \geq 2$ means can be gleaned from the observation that if we were to allow $\theta_1 = \frac{\pi}{2}$ and $\theta_2 = \frac{\pi}{2}$, then on the quadrant $\alpha = 2$ would correspond to the directions of reflection pointing flat along the sides of the wedge towards the corner, forcing the process to absorb there.

Further properties of reflected Brownian motions in a wedge have been studied in Varadhan and Williams [75], Williams [79, 80, 81]. These include results on hitting the corner of the wedge, recurrence classification, invariant measure, semimartingale representation and excursion decomposition. Some of these are indicated in Figure 8.4.

Using a localized version of Theorem 8.9, we can give a criterion for the existence and uniqueness of a reflected Brownian motion in a polygonal domain where the directions of reflection are constant along each side of the polygon. Namely, $\alpha < 2$ should be satisfied at each corner. A formula for the stationary distribution of such a reflected Brownian motion in a bounded polygon has been obtained by Harrison, Landau and Shepp [39].

Further reading. For reflected Brownian motions in polyhedral domains (RBM's) in three and more dimensions, a variety of open problems remain to be resolved. Most of the known results concern the problems of constructing RBM's with given geometric data and of determining stationary distributions for these processes. Path-by-path constructions for certain classes of RBM's have been given by Harrison and Reiman [41] and Lions and Sznitman [55]. For the processes constructed by Harrison and Reiman [41], a necessary and sufficient condition for the existence of a stationary distribution was given in [43] and product form stationary distributions were also discussed there.

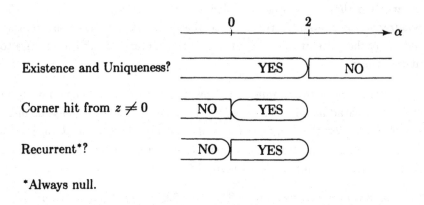

Figure 8.4.

8.6 Alternative Derivation of Equation (8.7)

We now give the promised alternative derivation of (8.7), by means of a brute force calculation due to McKean [57, pp. 75–76].

Notation. For the remainder of this chapter, let B denote a Brownian motion in $\mathbb{R}$ with initial value zero. Because of its mnemonic value we shall use M_t to denote $\max_{0 \leq s \leq t} B_s$. This should not be confused with the previous use of M to denote a martingale. For each t let X_t denote the non-negative random variable $M_t - B_t$.

It can be verified by direct calculation, using the known joint distribution of B_t and M_t, that X is equivalent in law to $|B|$. Freedman [33, pp. 32–34] has a neat proof of this. However, this result is also an immediate consequence of the paragraph following (8.6)—replace $\hat{B}$ (when $\hat{B}_0 = 0$) and Z there by $-B$ and X respectively. When the same substitutions are made in (8.7), the result is the following theorem. The alternate proof given below is purely computational.

Theorem 8.10. *For each* t,

$$(8.27) \qquad M_t = \lim_{\varepsilon \downarrow 0} \frac{1}{2\varepsilon} \lambda\{0 \le s \le t : X_s \le \varepsilon\}$$

where the limit is in L^2.

Proof. For each t and $\varepsilon > 0$ let $L_t^\varepsilon = (2\varepsilon)^{-1}\lambda\{0 \le s \le t : X_s \le \varepsilon\}$. We shall prove that $\overline{\lim}_{\varepsilon \downarrow 0} E\{(M_t - L_t^\varepsilon)^2\} = 0$. The following three densities are needed for this. The density of M_s and the joint density of B_s and M_s can be obtained using the reflection principle (e.g., see Karlin and Taylor [50, p. 346, 350]). They are given by

$$(8.28) \qquad P\{M_s \in dy\} = \frac{2}{\sqrt{2\pi s}} \exp\left(-\frac{y^2}{2s}\right) dy \quad \text{for} \quad y \ge 0, \quad \text{and}$$

$$(8.29) \quad P\{B_s \in dx, M_s \in dy\} = \sqrt{\frac{2}{\pi s^3}}(2y - x)\exp\left(-\frac{(2y - x)^2}{2s}\right) dx\, dy$$

for $x \le y$, and $y \ge 0$. Since X is equivalent in law to $|B|$, then for $0 \le r < s$, $x \ge 0$, and $y \ge 0$:

$$P\{X_r \in dx, X_s \in dy\}$$

$$(8.30) \qquad = \sqrt{\frac{2}{\pi r}} \exp\left(-\frac{x^2}{2r}\right) dx \frac{1}{\sqrt{2\pi(s - r)}}$$

$$\cdot \left\{\exp\left(-\frac{(y - x)^2}{2(s - r)}\right) + \exp\left(-\frac{(y + x)^2}{2(s - r)}\right)\right\} dy.$$

Now,

$$E\left\{(M_t - L_t^\varepsilon)^2\right\} = E\left\{(M_t)^2\right\} - 2E\left\{L_t^\varepsilon M_t\right\} + E\left\{(L_t^\varepsilon)^2\right\},$$

where by (8.28) we have

$$E\left\{(M_t)^2\right\} = \frac{2}{\sqrt{2\pi t}}\int_0^\infty y^2 e^{-y^2/2t}\, dy = E\left\{(B_t)^2\right\} = t.$$

From the definition of L_t^ε and symmetry we have

$$E\left\{(L_t^\varepsilon)^2\right\} = \frac{1}{4\varepsilon^2}E\left\{\left(\int_0^t 1_{[0,\varepsilon]}(X_s)\,ds\right)^2\right\}$$

$$= \frac{2}{4\varepsilon^2}E\left\{\int_0^t ds\int_0^s dr1_{[0,\varepsilon]}(X_r)1_{[0,\varepsilon]}(X_s)\right\}.$$

By Fubini's theorem and (8.30), the last line above equals

$$(8.31)\qquad \frac{1}{2\varepsilon^2}\int_0^t ds\int_0^s dr\int_0^\varepsilon dx\int_0^\varepsilon dy\frac{1}{\pi\sqrt{r(s-r)}}g(x,y,r,s)$$

where

$$g(x,y,r,s) \equiv e^{-x^2/2r}\left\{e^{-(y-x)^2/2(s-r)} + e^{-(y+x)^2/2(s-r)}\right\}.$$

As $\varepsilon \downarrow 0$, (8.31) converges to

$$\frac{1}{\pi}\int_0^t ds\int_0^s dr\frac{1}{\sqrt{r(s-r)}} = t.$$

By the definition of L_t^ε and Fubini's theorem, we have

$$(8.32)\qquad 2E\left\{L_t^\varepsilon M_t\right\} = \frac{1}{\varepsilon}\int_0^t dsE\left\{1_{[0,\varepsilon]}(X_s)M_t\right\}.$$

Since

$$M_t \geq \max_{s\leq r\leq t} B_r = B_s + \max_{s\leq r\leq t}(B_r - B_s),$$

the right side of (8.32) dominates

$$(8.33)\qquad \frac{1}{\varepsilon}\int_0^t dsE\left[1_{[0,\varepsilon]}(X_s)\left(B_s + E\left\{\max_{s\leq r\leq t}(B_r - B_s)|\mathcal{F}_s\right\}\right)\right].$$

Since B has stationary independent increments, it follows that

$$E\left\{\max_{s\leq r\leq t}(B_r-B_s)\mid \mathcal{F}_s\right\} = E\{M_{t-s}\} = \sqrt{\frac{2(t-s)}{\pi}}.$$

Here the second equality follows from (8.28). By the definition of X_s,

$$\{0\leq X_s\leq \varepsilon\} = \{M_s-\varepsilon\leq B_s\leq M_s\}.$$

Hence, using (8.29) we see that the expression (8.33) equals

$$\int_0^t ds\sqrt{\frac{2}{\pi s^3}}\int_0^\infty dy\,\varepsilon^{-1}\int_{y-\varepsilon}^y dx(2y-x)e^{-(2y-x)^2/2s}\left(x+\sqrt{\frac{2(t-s)}{\pi}}\right).$$

The limit as $\varepsilon\downarrow 0$ of the above equals:

$$\int_0^t ds\sqrt{\frac{2}{\pi s^3}}\int_0^\infty dy\,ye^{-y^2/2s}\left(y+\sqrt{\frac{2(t-s)}{\pi}}\right) = \int_0^t ds\left\{1+\frac{2}{\pi}\sqrt{\frac{t-s}{s}}\right\} = 2t.$$

By combining the above computations we obtain:

$$\varlimsup_{\varepsilon\downarrow 0} E\left\{(M_t-L_t^\varepsilon)^2\right\} \leq t-2t+t = 0. \quad \blacksquare$$

8.7 Exercises

1. For B, $\hat{B}$, Z and L as in Section 8.2, we have $(|B|, L) = (Z, L) = (\hat{B}+M, M)$ where $M_t = \max_{0\leq s\leq t}\hat{B}_s^-$. Use this together with the symmetry of Brownian motion and (8.29) to find the joint distribution of (B_t, L_t) for $t > 0$, given $B(0) = 0$ a.s.
(*Answer*:

$$P^0(B_t\in dx, L_t\in dy) = \frac{(|x|+y)}{\sqrt{2\pi t^3}}\exp\left\{\frac{-(|x|+y)^2}{2t}\right\}dx\,dy\ \text{for}\ x\in\mathbb{R}, y\geq 0.)$$

2. For B, L, $\hat{B}$ and M as in Exercise 1, use (8.29) and the fact that $(|B|, L) = (\hat{B}+M, M)$ to show that for $B(0) = 0$ a.s. and $t\geq 0$, $(|B|+$

$L)(t)$ has the same distribution as $|\mathbf{B}|(t)$, where $\mathbf{B}$ is a three-dimensional Brownian motion starting from the origin. In words this says that starting from the origin, $|B| + L$ has the same one-dimensional distributions as $|\mathbf{B}|$, the Bessel process with parameter $\frac{1}{2}$ started from the origin (cf. Chapter 5). In fact, these two processes are equivalent in law, i.e., have the same finite dimensional distributions. This is a consequence of the fact that $|B| + L = 2M - (-\hat{B})$ and the non-trivial result first proved by Pitman [62] using a random walk approximation that $2M - (-\hat{B})$ is equivalent in law to $|\mathbf{B}|$. For further discussion and developments of this so-called $2M - X$ property see [66] and [67].

3. Consider the function f defined in polar coordinates by

$$f(r, \theta) = \begin{cases} r^\alpha \cos(\alpha\theta - \theta_1), & \alpha \neq 0 \\ \ln r + \theta \tan \theta_1, & \alpha = 0 . \end{cases}$$

For the process Z described in Example 2 of Section 8.4, let Q^z be a probability measure under which $\mathbf{B}$ is a Brownian motion starting from $z \in S$ and define $\tau_0 = \inf\{t \geq 0 : \mathbf{Z}(t) = 0\}$. Use f, with $\alpha = \frac{1}{2}$, together with Itô's formula, to prove that for each $z \in S\backslash\{0\}$, $Q^z(\tau_0 < \infty) = 1$.

Hint: First verify that f is harmonic on $S\backslash\{0\}$ and satisfies $\nabla f \cdot v_j = 0$ on $\partial S_j \backslash\{0\}$, $j = 1, 2$.

Remark. The function f, together with a suitable representation for $z(\cdot)$ can be used to determine when the process associated with a solution of the submartingale problem described in Theorem 8.9 reaches the origin (see [75]).

GENERALIZED ITO FORMULA, CHANGE OF TIME AND MEASURE

9.1 Introduction

In this chapter we shall first obtain a generalized Itô formula for convex functions of Brownian motion. Then we shall prove a result which shows that Brownian motion is truly the canonical example of a continuous local martingale. Namely, if M is a continuous local martingale with quadratic variation $[M]_t$, then there is a random change of time τ_t such that $\{M_{\tau_t}, t \in I\!\!R_+\}$ is a Brownian motion up to the (random) time $[M]_\infty = \sup_{t \geq 0}[M]_t$. An application of this result shows that for a one-dimensional Brownian motion B starting from $x \geq 0$, there is a time change τ_t such that $\{B_{\tau_t}, t \in I\!\!R_+\}$ is equivalent in law to $|B|$. Finally, we show how local martingales behave under mutually absolutely continuous changes of probability measure. Using this, we obtain a formula for transforming a local martingale into a local martingale plus a state-dependent drift. We illustrate how this can be applied to obtain weak solutions of some stochastic differential equations.

9.2 Generalized Itô Formula

The following definitions and results concerning convex functions are well known.

Definitions. A real-valued function f defined on $\mathbb{R}$ is called *convex* if for each x and y in $\mathbb{R}$ and $\lambda \in [0, 1]$:

(9.1) $$f\left(\lambda x + (1 - \lambda)y\right) \leq \lambda f(x) + (1 - \lambda)f(y).$$

A weaker form of (9.1) is often used, namely

(9.2) $$f\left(\frac{1}{2}(x + y)\right) \leq \frac{1}{2}f(x) + \frac{1}{2}f(y).$$

It is known that (9.2), together with the measurability of f, implies (9.1), see e.g., Roberts and Varberg [65, Chapter VII]. We say f has a *right-hand derivative* at x if $\lim_{h \downarrow 0} \left(f(x + h) - f(x)\right)/h$ exists and is finite; in this case we denote the limit by $D^+ f(x)$. Similarly, if $\lim_{h \downarrow 0} \left(f(x) - f(x - h)\right)/h$ exists and is finite, it is called the *left-hand derivative* of f at x and we denote it by $D^- f(x)$. If $D^+ f$ and $D^- f$ are defined and are equal at x, then f is differentiable at x and the common value of $D^+ f(x)$ and $D^- f(x)$ is denoted by $f'(x)$ as usual.

Lemma 9.1. *A convex function f is continuous. Moreover, $D^+ f$ and $D^- f$ are defined everywhere on $\mathbb{R}$, they are both increasing functions, $D^+ f$ is right continuous and $D^- f$ is left continuous. Furthermore, $D^+ f$ equals $D^- f$ except on a countable set, and for all $a \leq b$ in $\mathbb{R}$:*

(9.3) $$f(b) - f(a) = \int_a^b D^+ f(x)\, dx = \int_a^b D^- f(x)\, dx = \int_a^b f'(x)\, dx.$$

Remark. The last integral in (9.3) is well-defined because $f'(x) = D^+ f(x)$ λ-a.e.

For a proof of this lemma, see e.g., loc. cit. pp. 3–7.

Let f be a convex function. Since D^+f is increasing, it induces a regular Borel measure μ on $\mathbb{R}$, defined by

$$(9.4) \qquad \mu((a,b]) = D^+f(b) - D^+f(a) \quad \text{for all} \quad a < b \quad \text{in} \quad \mathbb{R}.$$

Notation. Let C_c denote the space of real-valued continuous functions defined on $\mathbb{R}$ which have compact support. For each $n \in \mathbb{N}$, let C_c^n denote $C_c \cap C^n$, the space of n-times continuously differentiable functions having compact support.

For any $g \in C_c$, $\int_{-\infty}^{\infty} g \, d\mu$ is finite, and if $g \in C_c^2$, then using two integrations by parts and (9.3), we obtain

$$(9.5) \qquad \int_{-\infty}^{\infty} g \, d\mu = \int_{-\infty}^{\infty} g''(x)f(x) \, dx.$$

Let B denote a Brownian motion in $\mathbb{R}$ and let $L(t,x)$ denote the local time of B at x up to time t, defined by (7.7). The following theorem is due to Wang [76].

Theorem 9.2. *Let f be a convex function and μ be the measure on $\mathbb{R}$ associated with D^+f, which is defined by (9.4). Then for each t we have a.s.:*

$$(9.6) \qquad f(B_t) - f(B_0) = \int_0^t D^+f(B_s) \, dB_s + \frac{1}{2} \int_{-\infty}^{\infty} L(t,x) \, d\mu(x).$$

Proof. We apply Itô's formula to smooth approximations to f. Let $\{\phi_n\}$ be a sequence of positive C^∞ functions with compact supports shrinking to zero whose integrals over $\mathbb{R}$ equal 1, as in the paragraph before equation (7.3). Then $f_n \equiv \phi_n * f$ is in C^∞ and by Itô's formula we have a.s.:

$$(9.7) \qquad f_n(B_t) - f_n(B_0) = \int_0^t f_n'(B_s) \, dB_s + \frac{1}{2} \int_0^t f_n''(B_s) \, ds.$$

Since f is continuous, f_n converges to f uniformly on each compact subset of $\mathbb{R}$. Moreover, $f_n' = \phi_n * D^+ f$ converges to $D^+ f$ at all points of continuity of $D^+ f$, which is everywhere except possibly at a countable subset of $\mathbb{R}$, since $D^+ f$ is increasing. It follows by Fubini's theorem that $f_n'(B(s,\omega))$ converges to $D^+ f(B(s,\omega))$ for $(\lambda \times P)$-almost every $(s,\omega) \in \mathbb{R}_+ \times \Omega$. For each $k \in \mathbb{N}$, let

$$\tau_k = \inf\{s \geq 0 : |B_s| > k\}.$$

Then $|B_s| \leq k$ for $s \in (0, t \wedge \tau_k]$. Since $D^+ f$ is bounded on compact sets, $\{f_n'\}$ is uniformly bounded on $[-k, k]$. It follows by bounded convergence that

$$\lim_{n \to \infty} E\left\{ \int_0^{t \wedge \tau_k} \left| f_n'(B(s,\omega)) - D^+ f(B(s,\omega)) \right|^2 ds \right\} = 0.$$

Thus by the isometry (equation (2.10)), $\int_0^{t \wedge \tau_k} f_n'(B_s) \, dB_s$ converges to $\int_0^{t \wedge \tau_k} D^+ f(B_s) \, dB_s$ in L^2. Hence $\int_0^t f_n'(B_s) \, dB_s$ converges in pr. to $\int_0^t D^+ f(B_s) \, dB_s$.

Finally, by Corollary 7.4 we have a.s.:

$$(9.8) \qquad \int_0^t f_n''(B_s) \, ds = \int_{-\infty}^{\infty} L(t, x) f_n''(x) \, dx.$$

For each $g \in C_c^2$, as $n \to \infty$:

$$\int_{-\infty}^{\infty} g(x) f_n''(x) \, dx = \int_{-\infty}^{\infty} g''(x) f_n(x) \, dx$$

$$\to \int_{-\infty}^{\infty} g''(x) f(x) \, dx = \int_{-\infty}^{\infty} g(x) \, d\mu(x),$$

i.e., $f_n'' dx$ converges vaguely to $d\mu$. Since $x \to L(t, x)(\omega)$ is continuous with compact support for almost every ω, it follows from this vague convergence that the right member of (9.8) converges a.s. to $\int_{-\infty}^{\infty} L(t, x) \, d\mu(x)$.

By the limits established above, (9.6) follows by letting $n \to \infty$ in (9.7).

∎

Remarks.

1. If f is a differentiable function (not necessarily convex) such that f' is absolutely continuous, then f'' exists λ-a.e. and is integrable on each compact set. For such an f, by a similar argument to the above with f' and $f''dx$ respectively in place of $D^+ f$ and $d\mu$, it follows that (9.6) holds with these replacements. Rewriting the last term there, using Corollary 7.4, we obtain

$$f(B_t) - f(B_0) = \int\limits_0^t f'(B_s)\, dB_s + \frac{1}{2} \int\limits_0^t f''(B_s)\, ds,$$

the usual Itô formula.

2. Suppose $f(x) = |x|$, $D^+ f(x) = -1$ for $x < 0$ and $+1$ for $x \geq 0$, and $d\mu = 2\delta_0$ where δ_0 is the Dirac delta function. Then by substituting in (9.6) we obtain

$$|B_t| = |B_0| + \int\limits_0^t \operatorname{sgn}(B_s)\, dB_s + \int\limits_0^t 1_{\{0\}}(B_s)\, dB_s + L(t,0).$$

This is equivalent to formula (7.15) with $x = 0$ there, since the second stochastic integral in the above is zero a.s.

9.3 Change of Time

In this section we prove that any continuous local martingale M can be time-changed to a Brownian motion run up to the time $[M]_\infty$. Since a time-changed local martingale is adapted to a time-changed filtration, in the following we shall indicate explicitly the filtrations associated with all local martingales.

Let $\{M_t,\ \mathcal{F}_t,\ t \geq 0\}$ be a continuous local martingale on the probability space $(\Omega,\ \mathcal{F},\ P)$. Define

$$[M]_\infty \equiv \sup_{t \geq 0} [M]_t.$$

Then, since $[M]$ is P-a.s. increasing, $[M]_\infty = \lim_{t\to\infty}[M]_t$ P-a.s. For each $t \in \mathbb{R}_+$, let

$$\tau_t = \inf\{s \geq 0 : [M]_s > t\}.$$

Recall that by convention $\inf \emptyset = +\infty$.

We first consider the case where $[M]_\infty = \infty$ P-a.s. In this case, M can be time-changed to a Brownian motion run for all time.

Theorem 9.3. *For the continuous local martingale $\{M_t, \mathcal{F}_t, t \in \mathbb{R}_+\}$, suppose that $[M]_\infty = \infty$ P-a.s. Then $M^\tau = \{M_{\tau_t}, t \in \mathbb{R}_+\}$ is indistinguishable from a Brownian motion in $\mathbb{R}$.*

Proof. Since $M - M_0$ has the same quadratic variation as M, and the above properties of M and M^τ hold if and only if they hold for $M - M_0$ and $M^\tau - M_0$, we may suppose $M_0 = 0$. Also, by changing the definition of M on a set of probability zero, if necessary, we may suppose that $t \to [M]_t$ is increasing and $\lim_{t\to\infty}[M]_t = \infty$ everywhere on Ω.

As a function of t, τ_t is the right continuous inverse of $[M]_t$, and for each t, τ_t is an optional time. The condition $\lim_{t\to\infty}[M]_t = \infty$ ensures that τ_t is finite-valued for each t. Since M is continuous and $t \to \tau_t$ is right continuous, M^τ is right continuous, by composition. Moreover, $[M]_{\tau_t} = t$ for each t.

We use the characterization of Brownian motion given in Theorem 6.1 to prove the above theorem. First we show that $\{M_{\tau_t}, \mathcal{F}_{\tau_t}, t \in \mathbb{R}_+\}$ is an L^2-martingale. Let $\{\sigma_k, k \in \mathbb{N}\}$ be the localizing sequence for M defined by

$$\sigma_k = \inf\{t \geq 0 : |M_t| \geq k\}.$$

Then $M^k = \{M_{t\wedge\sigma_k}, \mathcal{F}_t, t \in \mathbb{R}_+\}$ is a bounded martingale and by Theorem 1.6, $\{M_{\tau_t\wedge\sigma_k}, \mathcal{F}_{\tau_t}, t \in \mathbb{R}_+\}$ is a bounded martingale. Since $M_0 = 0$, we have by (4.1),

$$(9.9) \qquad (M_{\tau_t\wedge\sigma_k})^2 = 2 \int_0^{\tau_t\wedge\sigma_k} M_s\, dM_s + [M]_{\tau_t\wedge\sigma_k}.$$

Since $M_{.\wedge\sigma_k}$ is bounded, it follows from the isometry (2.10) that the martingale $\left\{\int_0^{t\wedge\sigma_k} M_s\,dM_s, \mathcal{F}_t, t\in I\!\!R_+\right\}$ is L^2-bounded. Then by Theorem 1.6, $\left\{\int_0^{\tau_t\wedge\sigma_k} M_s\,dM_s, \mathcal{F}_{\tau_t}, t\in I\!\!R_+\right\}$ is an L^2-martingale. Hence the stochastic integral in (9.9) has zero expectation. Thus for all k,

$$(9.10) \qquad E\left\{(M_{\tau_t\wedge\sigma_k})^2\right\} = E\left\{[M]_{\tau_t\wedge\sigma_k}\right\} \le E\left\{[M]_{\tau_t}\right\} = t.$$

By the same kind of reasoning as in the proof of Theorem 6.1, $\{M_{\tau_t}, \mathcal{F}_{\tau_t}, t\in I\!\!R_+\}$ is an L^2-martingale.

Next we prove that almost all paths of M^τ are continuous. Since M^τ is right continuous, the set of ω for which it is discontinuous is given by

$$(9.11) \qquad \left\{\tau_{t-}\ne\tau_t \text{ and } M_{\tau_{t-}}\ne M_{\tau_t} \text{ for some } t>0\right\}.$$

Since $[M]_{\tau_{t-}} = [M]_{\tau_t} = t$, $[M]$ is increasing, and M is continuous, it follows that the above set is contained in:

$$\{\tau_{t-} < r < s < \tau_t,\ [M]_r = [M]_s \text{ and } M_r\ne M_s,$$
$$\text{for some } t>0 \text{ and rationals } r \text{ and } s\}.$$

The last set is contained in

$$\bigcup_{r,s}\{[M]_r = [M]_s\ ,\ M_r\ne M_s\}$$

where the union is over all rationals r, s such that $0 < r < s$. It suffices to prove that each set in this union has probability zero. For fixed r and s, let σ denote the first point of increase of $[M]$ after r, i.e., $\sigma = \inf\{u\ge r : [M]_u > [M]_r\}$. Then,

$$(9.12) \qquad \{[M]_r = [M]_s, M_r\ne M_s\} = \{\sigma\ge s, M_r\ne M_{\sigma\wedge s}\}.$$

By the definition of σ, $[M]_{\sigma\wedge s\wedge\sigma_k} - [M]_{r\wedge\sigma_k} = 0$. Thus by (4.1) we have

$$(9.13) \qquad (M_{\sigma\wedge s\wedge\sigma_k})^2 - (M_{r\wedge\sigma_k})^2 = 2\int_{r\wedge\sigma_k}^{\sigma\wedge s\wedge\sigma_k} M_u\,dM_u.$$

Moreover, since $\{\int_0^{t\wedge\sigma_k} M_u\,dM_u, \mathcal{F}_t, t\in I\!\!R_+\}$ is a martingale and σ is an optional time with respect to $\{\mathcal{F}_t\}$, it follows that the right member of

(9.13) has expectation zero. Since $M_{\sigma_s \wedge \sigma_k} - M_{r \wedge \sigma_k}$ is orthogonal to any $\mathcal{F}_{r \wedge \sigma_k}$-measurable random variable, the expectation of the left member of (9.13) is equal to that of $(M_{\sigma_s \wedge \sigma_k} - M_{r \wedge \sigma_k})^2$. Consequently, by taking the expectation of (9.13), letting $k \to \infty$, and using Fatou's lemma, we obtain $E\{(M_{\sigma_s} - M_r)^2\} = 0$. This implies that the right member of (9.12), and hence the left member, has probability zero, as desired. Thus, almost all paths of M^τ are continuous. By redefining M^τ on a P-null set so that it is continuous everywhere, we may suppose that $\{M_{\tau_t}, \mathcal{F}_{\tau_t}, t \in \mathbb{R}_+\}$ is a continuous L^2-martingale. The filtration $\{\mathcal{F}_{\tau_t}\}$ is automatically right continuous because $\{\mathcal{F}_t\}$ is assumed to be so and τ_t is non-decreasing and right continuous in t (cf. Exercise 1).

It remains to show that M^τ has quadratic variation given by $[M^\tau]_t = t$. For this we use the uniqueness of the decomposition of $(M^\tau)^2$ given in Theorem 4.6. By (4.1), with τ_t in place of t, we have

$$(9.14) \qquad (M_{\tau_t})^2 = 2 \int_0^{\tau_t} M_u \, dM_u + t.$$

We shall prove $\{\int_0^{\tau_t} M_u \, dM_u, \mathcal{F}_{\tau_t}, t \in \mathbb{R}_+\}$ is a local martingale by exhibiting a localizing sequence for it. For each k,

$$t_k \equiv \inf \{t \geq 0 : \tau_t \geq \sigma_k\},$$

is an optional time with respect to $\{\mathcal{F}_{\tau_t}, t \in \mathbb{R}_+\}$. We claim that

$$(9.15) \qquad \int_0^{\tau_t \wedge \sigma_k} M_u \, dM_u = \int_0^{\tau_{t \wedge t_k}} M_u \, dM_u \quad \text{for all } t \geq 0.$$

For the proof of this, note that

$$\tau_{(t \wedge t_k)-} \leq \tau_t \wedge \sigma_k \leq \tau_{t \wedge t_k}.$$

and $[M]_s = t \wedge t_k$ for $\tau_{(t \wedge t_k)-} \leq s \leq \tau_{t \wedge t_k}$. Thus, by the continuity of $s \to \int_0^s M_u \, dM_u$, the set on which (9.15) fails to hold is contained in

$$\{\tau_{(t \wedge t_k)-} \leq r < s \leq \tau_{t \wedge t_k}, [M]_r = [M]_s \text{ and } \int_0^r M_u dM_u \neq \int_0^s M_u dM_u,$$

$$\text{for some } t > 0 \text{ and rationals } r \text{ and } s\}.$$

But, by the decomposition formula (4.1) for $(M)^2$, the above is contained in

$$\bigcup_{r,s} \{[M]_r = [M]_s,\, M_r \neq M_s\},$$

where r and s are rationals such that $0 < r < s < \infty$. We have already shown that this last set has probability zero. Thus, (9.15) holds. Now, by the argument following (9.9), the left member of (9.15) defines an L^2 martingale with respect to $\{\mathcal{F}_{\tau_t}, t \geq 0\}$, and hence so does the right member. It follows that $\{\int_0^{\tau_t} M_u \, dM_u, \mathcal{F}_{\tau_t}, t \in \mathbb{R}_+\}$ is a local martingale with localizing sequence $\{t_k, k \in \mathbb{N}\}$. Then, by the uniqueness of the decomposition (9.14), we conclude that $[M^\tau]_t = t$.

We have proved that $\{M_{\tau_t}, \mathcal{F}_{\tau_t}, t \in \mathbb{R}_+\}$ is indistinguishable from a continuous L^2-martingale with quadratic variation at time t equal to t. The desired result then follows from Theorem 6.1. ∎

We now consider the general case where $[M]_\infty$ may be finite with positive probability. For this we introduce a Brownian motion independent of M which will be used to continue the time-changed version of M to a Brownian motion run for all time. Let $\tilde{B}$ be a one-dimensional Brownian motion defined on a probability space $(\tilde{\Omega}, \tilde{\mathcal{F}}, \tilde{P})$ that is independent of $(\Omega, \mathcal{F}, P)$ and suppose $\tilde{B}(0) = 0$. Define $(\bar{\Omega}, \bar{\mathcal{F}}, \bar{P})$ to be the completion of $(\Omega \times \tilde{\Omega}, \mathcal{F} \times \tilde{\mathcal{F}}, P \times \tilde{P})$. In the following, two sets are almost surely equal if their symmetric difference has probability zero.

Theorem 9.4. *For the continuous local martingale $\{M_t, \mathcal{F}_t, t \in \mathbb{R}_+\}$, we have P-a.s.*

(9.16)
$$\{\lim_{t \to \infty} M_t \text{ exists and is finite }\} = \{[M]_\infty < \infty\}$$

and

(9.17)
$$\limsup_{t \to \infty} M_t / (2[M]_t \log\log[M]_t)^{\frac{1}{2}} = 1 \quad \text{on } \{[M]_\infty = \infty\}.$$

Let

$$\Gamma \equiv \{[M]_\infty < \infty \text{ and } \lim_{t \to \infty} M_t \text{ does not exist in } \mathbb{R}\},$$

which is a P-null set by (9.16). Let $\tau_t = \inf\{s \geq 0 : [M]_s > t\}$ and define $M_\infty = \lim_{t\to\infty} M_t$, wherever this limit exists and is finite. For each $(t,\omega,\tilde{\omega}) \in \mathbb{R}_+ \times \Omega \times \tilde{\Omega}$, let

(9.18)
$$X_t(\omega,\tilde{\omega}) = \begin{cases} M_{\tau_t}(\omega) + (\tilde{B}_t(\tilde{\omega}) - \tilde{B}_{[M]_\infty(\omega)}(\tilde{\omega}))1_{\{[M]_\infty(\omega)<t\}} & \text{if } \omega \in \Omega\backslash\Gamma \\ 0 & \text{if } \omega \in \Gamma. \end{cases}$$

Then X on $(\bar{\Omega}, \bar{\mathcal{F}}, \bar{P})$ is indistinguishable from a Brownian motion in $\mathbb{R}$.

Note that for $\bar{P}$-a.e. $(\omega,\tilde{\omega}) \in \bar{\Omega}$, $X.(\omega,\tilde{\omega})$ is defined by the first expression in (9.18).

Proof. As in the proof of Theorem 9.3, we may and do suppose that $M_0 = 0$. We prove (9.16) first, using an argument adapted from Sharpe [70, Theorem 1.11]. For each positive integer k, define

$$\sigma_k = \inf\{t \geq 0 : |M_t| \geq k\}.$$

Then for each fixed k, $M^{\sigma_k} \equiv \{M_{t\wedge\sigma_k}, \mathcal{F}_{t\wedge\sigma_k}, t \in \mathbb{R}_+\}$ is a bounded martingale. Hence by the martingale convergence theorem (Theorem 1.5),

(9.19)
$$M_\infty^{\sigma_k} = \lim_{t\to\infty} M_{t\wedge\sigma_k}$$

exists and is finite P-a.s. In the following, for brevity, we omit the qualifier P-a.s. when it is clearly required. By Corollary 5.4 and Exercise 1 of Chapter 4,

$$E([M]_{\sigma_k}) = E([M^{\sigma_k}]_\infty) = E((M_\infty^{\sigma_k})^2) \leq k^2.$$

Thus, $[M]_{\sigma_k} < \infty$. Now, by the definition of σ_k and (9.19),

$$\{\lim_{t\to\infty} M_t \text{ exists and is finite }\} = \bigcup_k \{\sigma_k = \infty\}.$$

Combining the above we have

(9.20) $$\{\lim_{t\to\infty} M_t \text{ exists and is finite }\} \subset \{[M]_\infty < \infty\}.$$

To obtain the reverse inclusion, define

$$\tau_k = \inf\{t \geq 0 : [M]_t \geq k\},$$

for each positive integer k. Then, $M^{\tau_k} \equiv \{M_{t \wedge \tau_k}, \mathcal{F}_{t \wedge \tau_k}, t \in \mathbb{R}_+\}$ is a continuous local martingale, and by Corollary 5.4,

$$E([M^{\tau_k}]_\infty) = E([M]_{\tau_k}) \leq k.$$

Then by Exercise 2 of Chapter 4 and the martingale convergence theorem (Theorem 1.5), M^{τ_k} is an L^2-bounded martingale such that $\lim_{t\to\infty} M_t^{\tau_k} = \lim_{t\to\infty} M_{t \wedge \tau_k}$ exists and is finite. By combining this with

$$(9.21) \qquad \{[M]_\infty < \infty\} = \bigcup_k \{\tau_k = \infty\},$$

we obtain the reverse of the inclusion in (9.20), and then the equality (9.16) follows.

Next we prove that X is a Brownian motion on $(\bar{\Omega}, \bar{\mathcal{F}}, \bar{P})$ (cf. [34, pp. 292–293]). For each $t \in \mathbb{R}_+$, define

$$(9.22) \qquad Y_t = \begin{cases} M_{\tau_t} & \text{on} \quad \Omega \backslash \Gamma, \\ 0 & \text{on} \quad \Gamma. \end{cases}$$

By the continuity of $[M]$, for each $t \in \mathbb{R}_+$, $\{\tau_t < \infty\} = \{t < [M]_\infty\}$ and $[M]_{\tau_t} = t \wedge [M]_\infty$. By the definition of Γ, Y is well defined on Ω and is P-a.s. given by the upper expression in (9.22). A similar argument to that for Theorem 9.3 shows that $\{Y_t, \mathcal{F}_{\tau_t}, t \in \mathbb{R}_+\}$ is indistinguishable from a continuous L^2-martingale with quadratic variation: $[Y]_t = t \wedge [M]_\infty$. Let $\bar{\mathcal{F}}_t$ denote the augmentation of $\mathcal{F}_{\tau_t} \times \tilde{\mathcal{F}}_t$ by the $\bar{P}$-null sets in $\bar{\mathcal{F}}$. Then $\{\bar{\mathcal{F}}_t\}$ is a standard filtration (see [34; (7.5), p. 300]) and X defined by (9.18) is adapted to it. Moreover, Z, defined by

$$Z_t(\omega, \tilde{\omega}) = 1_{\{[M]_\infty(\omega) < t\}} \quad \text{for all} \quad (t, \omega, \tilde{\omega}) \in \mathbb{R}_+ \times \Omega \times \tilde{\Omega},$$

is a predictable process on $(\bar{\Omega}, \bar{\mathcal{F}}, \{\bar{\mathcal{F}}_t\})$ (cf. Theorem 3.1). Extend Y_t and $\tilde{B}_t$ to $\bar{\Omega}$ by setting $Y_t(\omega, \tilde{\omega}) = Y_t(\omega)$ and $\tilde{B}_t(\omega, \tilde{\omega}) = \tilde{B}_t(\tilde{\omega})$. Then we have $\bar{P}$-a.s. on $\bar{\Omega}$:

$$X_t = Y_t + \int_0^t Z_s d\tilde{B}_s \quad \text{for all} \quad t \geq 0.$$

It follows that $\{X_t, \bar{\mathcal{F}}_t, t \in \mathbb{R}_+\}$ is indistinguishable from a continuous local martingale on $(\bar{\Omega}, \bar{\mathcal{F}}, \bar{P})$. Since Y and $\tilde{B}$ are *independent* martingales on $(\bar{\Omega}, \bar{\mathcal{F}}, \bar{P})$, by Theorem 5.7 and Exercise 6 of Chapter 5, we have for each $t \geq 0$,

$$\left[Y, \int Z d\tilde{B} \right]_t = 0.$$

and hence

$$[X]_t \equiv [X, X]_t = [Y]_t + 2\left[Y, \int Z d\tilde{B} \right]_t + \left[\int Z d\tilde{B} \right]_t$$

$$= t \wedge [M]_\infty + \int_0^t 1_{\{[M]_\infty < s\}} ds$$

$$= t.$$

Thus, $\{X_t, \bar{\mathcal{F}}_t, t \in \mathbb{R}_+\}$ is indistinguishable from a continuous local martingale on $(\bar{\Omega}, \bar{\mathcal{F}}, \bar{P})$ with quadratic variation process $[X]_t = t$. It follows from the characterization given in Theorem 6.1 that X is indistinguishable from a Brownian motion on $(\bar{\Omega}, \bar{\mathcal{F}}, \bar{P})$.

Now the law of the iterated logarithm for Brownian motion states that (see, for example, [47]),

$$\limsup_{t \to \infty} X_t / (2t \log \log t)^{\frac{1}{2}} = 1 \quad \bar{P}\text{-a.s.}$$

Hence by (9.18),

$$(9.23) \qquad \limsup_{t \to \infty} M_{\tau_t} / (2t \log \log t)^{\frac{1}{2}} = 1 \quad P\text{-a.s. on} \quad \{[M]_\infty = \infty\}.$$

By the same proof as in Theorem 9.3, P-almost surely, M is constant on an interval if and only if $[M]$ is constant there. It follows that τ_t and t may be replaced by t and $[M]_t$, respectively, in the quotient in (9.23) to yield (9.17). ∎

We now combine Theorem 9.3 on time change with the decomposition in (7.5), to obtain a realization of reflected Brownian motion by deleting the time during which a given Brownian motion is negative. This corresponds to the intuitive idea of placing, end-to-end, the successive excursions of a Brownian motion above zero, as illustrated in Figures 9.1–9.2.

This result is originally due to Itô and McKean [57, p. 81], but the proof given below is due to El Karoui and Chaleyat-Maurel [28, Proposition 1.4.1].

For this let B denote a Brownian motion in $\mathbb{R}$ with $B_0 \geq 0$. For each t let $A_t = \int_0^t 1_{\{B_s \geq 0\}} ds$. Let $\{\tau_t, t \in \mathbb{R}_+\}$ denote the right continuous inverse of A, given by $\tau_t = \inf\{s \geq 0 : A_s > t\}$.

Theorem 9.5. $\{B_{\tau_t}, t \in \mathbb{R}_+\}$ *is a continuous process equivalent in law to* $\{|B_t|, t \in \mathbb{R}_+\}$.

Proof. Set $x = 0$ in (7.5). Since $B_0 \geq 0$ this gives

$$(9.24) \qquad B_t^+ = B_0 + \int_0^t 1_{\{B_s \geq 0\}} dB_s + \frac{1}{2} L_t$$

where $L_t = L(t, 0)$ is defined by (7.7). Let $Y_t = \int_0^t 1_{\{B_s \geq 0\}} dB_s$. Then $\{Y_t, \mathcal{F}_t, t \in \mathbb{R}_+\}$ is a continuous L^2-martingale, and by (5.21) we have a.s., $[Y]_t = \int_0^t 1_{\{B_s \geq 0\}} ds = A_t$. Since $\lim_{t \to \infty} A_t = \infty$ a.s., it follows by Theorem 9.3 that $\{Y_{\tau_t}, t \in \mathbb{R}_+\}$ is indistinguishable from a Brownian motion. Let $\hat{B}_t = B_0 + Y_{\tau_t}$. Replacing t by τ_t in (9.24) we obtain

$$(9.25) \qquad B_{\tau_t}^+ = \hat{B}_t + \frac{1}{2} L_{\tau_t}.$$

Since $A_{\tau_{t-}} = A_{\tau_t}$, when $\tau_{t-} \neq \tau_t$ we have $B_s \leq 0$ for all $s \in [\tau_{t-}, \tau_t]$ and consequently $B_{\tau_{t-}}^+ = B_{\tau_t}^+ = 0$. It follows that $t \to B_{\tau_t}^+$ is continuous. From its definition, τ_t is a point of increase of A, and hence $B_{\tau_t} \geq 0$. Thus, $B_{\tau_t} = B_{\tau_t}^+$ for all t. Hence, $t \to B_{\tau_t}$ is continuous and we may replace $B_{\tau_t}^+$ by B_{τ_t} in (9.25). Since $B_0 \geq 0$, it follows that a.s., $A_t > 0$ for all $t > 0$, and consequently $\tau_0 = 0$ and $L_{\tau_0} = 0$. Furthermore, since L can increase only when B is at zero, L_{τ_t} can increase only when $B_{\tau_t} = 0$. Thus when $B_{\tau_t}^+$ is replaced by B_{τ_t} in (9.25), the result is a decomposition which is a.s. of the form in Lemma 8.1 with $x_t = \hat{B}_t$, $z_t = B_{\tau_t}$ and $y_t = \frac{1}{2} L_{\tau_t}$. Now it follows from the uniqueness stated there that $B_{\tau_t} = \hat{B}_t + \max_{0 \leq s \leq t} \hat{B}_s^-$ a.s. for all t. Since $\hat{B}$ is indistinguishable from a Brownian motion, the paragraph following (8.6) shows that $\{B_{\tau_t}, t \in \mathbb{R}_+\}$ is equivalent in law to $|\hat{B}|$ and hence to $|B|$. ∎

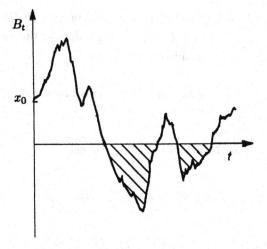

Figure 9.1.

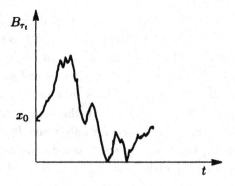

Figure 9.2.

9.4 Change of measure

In this section, we consider the behavior of local martingales under mutually absolutely continuous changes of probability measure. When applied to solutions of stochastic differential equations, these results can be used to change the drift, simply by changing the ambient probability measure in a mutually absolutely continuous manner over each "finite time horizon", i.e., on the σ-fields generated by the process on each finite time interval. The prototype of this formula, where the local martingale is a one-dimensional Brownian motion, was developed by Cameron and Martin [8, 9, 10]. Subsequently, Girsanov [36] generalized their transformation formula to a multi-dimensional Brownian motion and applied it to change the drifts of solutions of multi-dimensional stochastic differential equations. Maruyama [56] had foreseen some of the results of Girsanov [36]. Generalizations and refinements of the formula were subsequently made by a variety of authors. In deference to Girsanov's innovative use of the modern notation and terminology of Itô's theory of stochastic integration, the general change of measure formula is commonly referred to as the *Girsanov formula*. The term *Cameron-Martin formula* is usually reserved for the special case where the local martingale is a Brownian motion. We shall follow this custom. The reader should be warned that there is no common agreement on this however, for example, some authors use the blanket term Cameron-Martin-Girsanov formula to cover all cases. Before proceeding to the discussion of the general transformation result (Theorem 9.8), we give a simple application for the case of a one-dimensional Brownian motion and a bounded drift.

Definition. Two measures defined on the same measurable space are said to be *equivalent* if and only if they are *mutually absolutely continuous*.

Notation. In the following, we will frequently be dealing with more than one probability measure at a time. Whenever the meaning is not clear from the context, we shall use a superscript on the expectation operator to indicate the measure under which the expectation is to be taken.

Example. *(Brownian motion plus a bounded drift.)* Suppose X is a one-dimensional Brownian motion on $(\Omega, \mathcal{F}, P)$ and $b : I\!R \to I\!R$ is a bounded measurable function. Then, since b is bounded, $\int_0^t b(X_s) \, dX_s$ defines a continuous L^2-martingale, and then by (6.18) and Theorem 6.2,

$$(9.26) \qquad \rho_t = \exp\left(\int_0^t b(X_s) \, dX_s - \frac{1}{2} \int_0^t b^2(X_s) \, ds\right)$$

defines a positive continuous L^2-martingale on $(\Omega, \mathcal{F}, P)$. For each $t \in I\!R_+$, let P_t denote the restriction of P to $\mathcal{F}_t$ and let Q_t be the probability measure on $\mathcal{F}_t$ that is absolutely continuous with respect to P_t and whose Radon-Nikodym derivative is given by

$$(9.27) \qquad \frac{dQ_t}{dP_t} = \rho_t \quad \text{on } \mathcal{F}_t.$$

Since $\rho_t > 0$ P-a.s., Q_t is equivalent to P_t, and by the martingale property of $\{\rho_t, \mathcal{F}_t, t \in I\!R_+\}$, the Q_t's are consistent (see (9.41)). Let

$$(9.28) \qquad B_s = X_s - \int_0^s b(X_u) \, du \quad \text{for each } s \in I\!R_+.$$

It will be shown later that for each $t \in I\!R_+$, $\{B_s, s \in [0,t]\}$ is a Brownian motion on the interval $[0,t]$ under Q_t. In other words, the probability measure has been changed on $\mathcal{F}_t$ in an absolutely continuous manner from P_t to Q_t so that X restricted to the time interval $[0,t]$ has changed from a Brownian motion under P_t to a Brownian motion plus a state-dependent drift under Q_t. If $\{\rho_t, \mathcal{F}_t, t \in I\!R_+\}$ is uniformly integrable, the Q_t's can be extended to a measure Q that is absolutely continuous with respect to P on $\mathcal{F}$. However, in general the uniform integrability does not hold, and there is no such extension (cf. Exercise 3). For example, the laws of Brownian motion, and Brownian motion plus a constant non-zero drift, on the time interval $[0,\infty)$ are mutually singular. Further discussion of the question of existence or non-existence of such an extension Q will be given later in the paragraph following (9.46). For the present, this question is not important because we shall be concerned only with events that are contained in some $\mathcal{F}_t$, for which the transformation given by (9.27) applies.

The above transformation can be used to obtain probabilistic representations for solutions of certain second order differential equations in which a first order derivative term is present. For example, fix $t \in \mathbb{R}_+$ and consider Itô's formula applied to $X_{\cdot \wedge t} = B_{\cdot \wedge t} + \int_0^{\cdot \wedge t} b(X_u)\, du$ on the complete probability space $(\Omega, \mathcal{F}_t, Q_t)$, and to a function f that is twice continuously differentiable on $\mathbb{R}$. Then, we have Q_t-a.s. for all $s \in \mathbb{R}_+$,

(9.29)
$$f(X_{s \wedge t}) - f(X_0)$$
$$= \int_0^{s \wedge t} f'(X_u)\, dB_u + \int_0^{s \wedge t} f'(X_u) b(X_u)\, du + \frac{1}{2} \int_0^{s \wedge t} f''(X_u)\, du.$$

If f satisfies the differential equation

$$\frac{1}{2} f'' + b f' = 0 \quad \text{in } \mathbb{R},$$

then the last two terms in (9.29) sum to zero, and we obtain Q_t-a.s. for all $s \in \mathbb{R}_+$,

(9.30)
$$f(X_{s \wedge t}) - f(X_0) = \int_0^{s \wedge t} f'(X_u)\, dB_u.$$

If f or f' is bounded on $\mathbb{R}$, then the right member of (9.30) defines a continuous martingale under Q_t. For if f is bounded, the stochastic integral, being equal to the left member of (9.30), defines a bounded local martingale and hence a bounded martingale under Q_t (cf. Proposition 1.8); if f' is bounded, the integrand is bounded and the stochastic integral defines an L^2-martingale (cf. Theorem 2.5). In either case, taking expectations with respect to Q_t in (9.30) yields

(9.31)
$$E^{Q_t}[f(X_s) - f(X_0)] = 0 \quad \text{for all } s \in [0, t].$$

For $x \in \mathbb{R}_+$, suppose P^x is a probability measure on $(\Omega, \mathcal{F})$ such that X is a Brownian motion under P^x and $X(0) = x$ P^x-a.s., and for each $t \in \mathbb{R}_+$, let Q_t^x denote the Q_t-measure corresponding to P^x. Since Q_t^x is absolutely continuous with respect to P^x on $\mathcal{F}_t$ and $\{X(0) = x\} \in \mathcal{F}_0 \subset \mathcal{F}_t$, we have $X(0) = x$ Q_t^x-a.s. Then, by (9.31) with $s = t$, and (9.27), we have

(9.32)
$$f(x) = E^{Q_t^x}[f(X_t)]$$
$$= E^{P^x}\left[f(X_t) \exp\left(\int_0^t b(X_s)\, dX_s - \frac{1}{2} \int_0^t b^2(X_s)\, ds \right) \right].$$

This is a form of the Cameron-Martin formula. By comparison, applying Itô's formula to X on $(\Omega, \mathcal{F}, P^x)$, we obtain P^x-a.s.,

$$f(X_t) - f(X_0) = \int_0^t f'(X_s)\,dX_s + \frac{1}{2}\int_0^t f''(X_s)\,ds$$

$$= \int_0^t f'(X_s)\,dX_s - \int_0^t b(X_s)f'(X_s)\,ds.$$

If f' is bounded on $\mathbb{R}$, then $\int_0^t f'(X_s)\,dX_s$ defines an L^2-martingale under P^x and taking expectations in the above yields

(9.33) $$f(x) = E^{P^x}[f(X_0)] = E^{P^x}\left[f(X_t) + \int_0^t b(X_s)f'(X_s)ds\right].$$

To see the advantage of the representation (9.32) over (9.33), consider the case where b is constant. Then the right member of (9.32) reduces to

$$E^{P^x}\left[f(X_t)\exp\left(b(X_t - X_0) - \frac{1}{2}b^2 t\right)\right],$$

which can be evaluated using the P^x-distribution of the Brownian motion X at the single time t, whereas the representation in (9.33) involves the P^x-distribution of X over the entire interval $[0, t]$.

We have indicated above that the change of measure formula (9.27) can be used as a quantitative tool for simplifying calculations involving Brownian motion plus a drift. It is also useful, and this is perhaps its predominant use, as a qualitative tool. For example, by the absolute continuity of Q_t with respect to P_t for each t, if one knows that, with positive probability, a certain event occurs in finite time for a Brownian motion, then this must also be true for a Brownian motion plus a bounded drift. Of course, the probabilities will not be the same in general, but they will both be positive. For example, one-dimensional Brownian motion hits the origin in finite time with probability one. Consequently, a Brownian motion plus a bounded drift will hit the origin in finite time with positive probability. However, the absolute continuity does not in general extend to events depending on the whole history of the path, such as those concerned with transience or recurrence. For example, while one-dimensional Brownian motion is null recurrent, one-dimensional Brownian motion plus a constant non-zero drift is transient.

There is a similar change of measure transformation for d-dimensional Brownian motion. This can be used to represent solutions of some elliptic partial differential equations. It is specified as follows. Suppose $X = (X^1, \ldots, X^d)$ is a d-dimensional Brownian motion on $(\Omega, \mathcal{F}, P)$ and $b : \mathbb{R}^d \to \mathbb{R}^d$ is a bounded Borel measurable function. For each $t \in \mathbb{R}_+$, define

$$\rho_t = \exp\left(\int_0^t b(X_s)dX_s - \frac{1}{2}\int_0^t |b(X_s)|^2 \, ds \right),$$

where $\int_0^t b(X_s)dX_s \equiv \sum_{i=1}^d b_i(X_s)dX_s^i$. Using this ρ_t, define Q_t to be absolutely continuous with respect to P_t on $\mathcal{F}_t$ so that (9.27) holds. Let $B = \{B_s, s \in \mathbb{R}_+\}$ be the d-dimensional process defined as in (9.28). Then on the probability space $(\Omega, \mathcal{F}_t, Q_t)$, $\{B_s, s \in [0, t]\}$ is a d-dimensional Brownian motion on the time interval $[0, t]$.

Local martingales under mutually absolutely continuous changes of probability measure.

We now study the behavior of local martingales under mutually absolutely continuous changes of probability measure. We shall apply this by showing how a probability measure can be changed on the σ-fields $\mathcal{F}_t$, $t \in \mathbb{R}_+$, to add a state-dependent "drift" to a local martingale. Our treatment is similar, though more expanded, than that in Durrett [27, §2.13]. Finally, we shall illustrate this in the special case when the local martingale is a Brownian motion. This will allow us to justify the claims made in the above example and also treat the case of unbounded drift, which was not discussed above.

As usual, $(\Omega, \mathcal{F}, P)$ is a complete probability space with a given standard filtration $\{\mathcal{F}_t, t \in \mathbb{R}_+\}$, and $\mathcal{F}_\infty \equiv \bigvee_{t \in \mathbb{R}_+} \mathcal{F}_t$.

For the following development, up to the end of the proof of Theorem 9.8, we suppose Q is a probability measure on $(\Omega, \mathcal{F})$ such that Q is equivalent to P. Let $\rho = dQ/dP$, the Radon-Nikodym derivative of Q with respect to P. For each $t \in \mathbb{R}_+$, define

$$\rho_t = E^P[\rho \,|\, \mathcal{F}_t].$$

Then $\{\rho_t, \mathcal{F}_t, t \in \mathbb{R}_+\}$ is a uniformly integrable martingale on $(\Omega, \mathcal{F}, P)$. We may suppose it is right continuous, since there is always a right continuous version of it [12, p. 29]. Moreover, $\rho_\infty \equiv E[\rho \,|\, \mathcal{F}_\infty]$ is P-a.s. equal to $\lim_{t\to\infty} \rho_t$, by the martingale convergence theorem. For some arguments involving localization, we shall need the following.

Lemma 9.6. *Suppose τ is an optional time. Let Q_τ and P_τ denote the restrictions of Q and P, respectively, to $\mathcal{F}_\tau$. Then*

$$\rho_\tau = \frac{dQ_\tau}{dP_\tau}.$$

Here $\rho_\tau(\omega) = \rho_{\tau(\omega)}(\omega)$.

Proof. By Doob's stopping theorem, $\rho_\tau = E^P[\rho \,|\, \mathcal{F}_\tau]$. Thus, for $F \in \mathcal{F}_\tau$,

$$E^{P_\tau}[\rho_\tau 1_F] = E^P[\rho_\tau 1_F] = E^P[\rho 1_F].$$

By the definition of ρ, $E^P[\rho 1_F] = Q(F)$, and since $F \in \mathcal{F}_\tau$, $Q(F) = Q_\tau(F)$. Combining the above, we obtain $E^{P_\tau}[\rho_\tau 1_F] = Q_\tau(F)$, which proves the desired result. ∎

Theorem 9.7. *Let $\{M_t, t \in \mathbb{R}_+\}$ be a right continuous stochastic process on $(\Omega, \mathcal{F})$ with $M_t \in \mathcal{F}_t$ for all $t \in \mathbb{R}_+$. Then, $\{\rho_t M_t, \mathcal{F}_t, t \in \mathbb{R}_+\}$ is a local martingale under P if and only if $\{M_t, \mathcal{F}_t, t \in \mathbb{R}_+\}$ is a local martingale under Q.*

Proof. Without loss of generality, we may and do assume that $M_0 = 0$, since $\{\rho_t M_0, \mathcal{F}_t, t \in \mathbb{R}_+\}$ is a P-local martingale and the random variable M_0 defines a Q-local martingale. For the proof of the "only if" part, suppose $\{\rho_t M_t, \mathcal{F}_t, t \in \mathbb{R}_+\}$ is a local martingale under P and $\{\tau_n\}$ is a localizing sequence for it, i.e., $\tau_n \uparrow \infty$ P-a.s. and for each n, $\{\rho_{t\wedge\tau_n} M_{t\wedge\tau_n}, \mathcal{F}_{t\wedge\tau_n}, t \in$

$\mathbb{R}_+\}$ is a martingale on $(\Omega, \mathcal{F}, P)$. Then, for any $0 \le s < t < \infty$ and $F \in \mathcal{F}_{s \wedge \tau_n}$, we have by Lemma 9.6,

$$\int_F M_{t \wedge \tau_n} dQ = \int_F M_{t \wedge \tau_n} dQ_{t \wedge \tau_n} = \int_F \rho_{t \wedge \tau_n} M_{t \wedge \tau_n} dP_{t \wedge \tau_n}.$$

By using $P = P_{t \wedge \tau_n}$ on $\mathcal{F}_{t \wedge \tau_n}$, and the martingale property of $\rho_{\cdot \wedge \tau_n} M_{\cdot \wedge \tau_n}$ under P, we see that the right member above equals,

$$\int_F \rho_{t \wedge \tau_n} M_{t \wedge \tau_n} dP = \int_F \rho_{s \wedge \tau_n} M_{s \wedge \tau_n} dP.$$

Then, using $P = P_{s \wedge \tau_n}$, $Q = Q_{s \wedge \tau_n}$ on $\mathcal{F}_{s \wedge \tau_n}$, and Lemma 9.6, we find that the right member above equals,

$$\int_F \rho_{s \wedge \tau_n} M_{s \wedge \tau_n} dP_{s \wedge \tau_n} = \int_F M_{s \wedge \tau_n} dQ_{s \wedge \tau_n} = \int_F M_{s \wedge \tau_n} dQ.$$

Hence, $\{M_{t \wedge \tau_n}, \mathcal{F}_{t \wedge \tau_n}, t \in \mathbb{R}_+\}$ is a martingale on $(\Omega, \mathcal{F}, Q)$. Since Q is equivalent to P, we have: $\tau_n \uparrow \infty$ Q-a.s. It follows that $\{M_t, \mathcal{F}_t, t \in \mathbb{R}_+\}$ is a local martingale under Q, with $\{\tau_n\}$ as a localizing sequence.

The "if" part is proved in a similar manner. ∎

For the next theorem, we need to assume that $\{\rho_t, \mathcal{F}_t, t \in \mathbb{R}_+\}$ has a continuous version. This assumption will be satisfied in the application and the Brownian motion example that follow the theorem. In general, such an assumption would be valid if every local martingale $N = \{N_t, \mathcal{F}_t, t \in \mathbb{R}_+\}$ on $(\Omega, \mathcal{F}, P)$ had a continuous version. Recall from the Proposition at the end of Section 2.3 that the latter holds if and only if every optional time is predictable. In particular, this holds if $\{\mathcal{F}_t\}$ is the standard filtration associated with a d-dimensional Brownian motion, $d \ge 1$.

The result of the following Theorem is often referred to as a *Girsanov transformation*.

Theorem 9.8. *Suppose $X = \{X_t, \mathcal{F}_t, t \in \mathbb{R}_+\}$ is a continuous local martingale under P. Let Q, ρ and $\{\rho_t, \mathcal{F}_t, t \in \mathbb{R}_+\}$ be as above. Suppose the martingale $\{\rho_t, \mathcal{F}_t, t \in \mathbb{R}_+\}$ has a continuous version. Using such a* *continuous version, let*

$$(9.34) \qquad A_t = \int_0^t \rho_s^{-1} \, d[\rho, X]_s,$$

wherever the right member is well defined and finite for all t. Then $A = \{A_t, t \in \mathbb{R}_+\}$ is well defined P-a.s. If A is defined to be identically zero on the remaining P-null set, then A is locally of bounded variation (under either P or Q) and $\{X_t - A_t, \mathcal{F}_t, t \in \mathbb{R}_+\}$ is a continuous local martingale under Q, with a quadratic variation process that is the same as that for X under P.

Remark. Thus, under the change of probability from P to Q, X changes from a continuous local martingale under P to a continuous semimartingale (the sum of a continuous local martingale and a continuous process that is locally of bounded variation) under Q.

Proof. Throughout this proof, by choosing a suitable version, we may and do assume that $\{\rho_t, \mathcal{F}_t, t \in \mathbb{R}_+\}$ is a continuous martingale under P. Combining this with the fact that X is a continuous local martingale under P, it follows that $[\rho, X]$ is well defined (cf. Section 5.3). We first prove that P-a.s., A_t is well defined and finite for all t. For each positive integer n, let $\tau_n = \inf\{t \geq 0 : \rho_t \leq 1/n\}$. Then P-a.s. on $\{t < \tau_n\}$:

$$(9.35) \qquad \int_0^t \rho_s^{-1} d\,|[\rho, X]|_s \leq n\,|[\rho, X]|_t \leq n([\rho]_t[X]_t)^{1/2} < \infty,$$

where $|[\rho, X]|_s$ denotes the variation of $[\rho, X]$ on $[0, s]$ and the second inequality holds P-a.s. by (5.11). Let $\tau = \sup_n \tau_n$. Since $\{\rho_t, \mathcal{F}_t, t \in [0, \infty]\}$ is a continuous martingale under P, by Doob's stopping theorem we have

$$
\begin{aligned}
(9.36) \qquad E^P[\rho; \tau_n < \infty] &= E^P[E^P[\rho \,|\, \mathcal{F}_{\tau_n}]; \tau_n < \infty] \\
&= E^P[\rho_{\tau_n}; \tau_n < \infty] \\
&\leq 1/n.
\end{aligned}
$$

Letting $n \to \infty$ in the above, we obtain

$$E^P[\rho; \tau < \infty] = 0.$$

By the equivalence of Q to P, ρ is P-a.s. strictly positive, and so by the above, $P(\tau < \infty) = 0$. It follows that P-a.s., A_t is well defined and finite for all $t \in I\!\!R_+$. If A is defined to be identically zero on the remaining P-null set, then A is a continuous adapted process which by (9.35) is locally of bounded variation.

Applying Itô's formula (5.22) with $f(x_1, x_2, y) = x_1(x_2 - y)$ for $x_1, x_2, y \in I\!\!R$, and $M_t^1 = \rho_t$, $M_t^2 = X_t$, and $V_t^1 = A_t$, we obtain P-a.s. for all $t \in I\!\!R_+$:

$$\rho_t(X_t - A_t) - \rho_0(X_0 - A_0)$$
$$= \int_0^t (X_s - A_s)d\rho_s + \int_0^t \rho_s dX_s - \int_0^t \rho_s dA_s + [\rho, X]_t.$$

Now A was defined so that the combination of the last two terms above is zero P-a.s. Then, by Theorem 2.11, $\{\rho_t(X_t - A_t), \mathcal{F}_t, t \in I\!\!R_+\}$ is a continuous local martingale under P. It follows from Theorem 9.7 that $X - A = \{X_t - A_t, \mathcal{F}_t, t \in I\!\!R_+\}$ is a local martingale under Q. Since Q is equivalent to P, it has the same null sets, and consequently the filtration $\{\mathcal{F}_t\}$ is still standard under Q. Hence, $X - A$ is a *continuous* local martingale under Q.

By Exercise 3 of Chapter 4, for each fixed $t \in I\!\!R_+$, the sums S_t^n defined there converge in Q-probability to the quadratic variation of $X - A$ at time t. On the other hand, since X is a continuous local martingale under P, along a subsequence, the same sums converge P-a.s. to $[X]_t$. Since Q is equivalent to P, this also holds Q-a.s. Then, by the uniqueness of limits, it follows that the quadratic variation $[X - A]_t$ of $X - A$ under Q is equal to the quadratic variation $[X]_t$ of the local martingale X under P. Since these quadratic variation processes are continuous, it follows that they are indistinguishable under Q and hence we regard them as the same. ∎

Application. We shall now apply the above result locally on each $\mathcal{F}_t$. For this, let $X = \{X_t, \mathcal{F}_t, t \in I\!\!R_+\}$ be a continuous local martingale on $(\Omega, \mathcal{F}, P)$ and $b : I\!\!R \to I\!\!R$ be a Borel measurable function. For each $t \in I\!\!R_+$ and suitably integrable b, we shall perform a mutually absolutely continuous change of probability measure on $\mathcal{F}_t$ under which $X_{\cdot \wedge t}$ is transformed to a local martingale plus a drift term of the form $\int_0^{\cdot \wedge t} b(X_s) d[X]_s$. By analogy with the example at the beginning of this section, replacing ds by $d[X]_s$ there, a good candidate for the Radon-Nikodym derivative governing this change of measure is

$$(9.37) \qquad \rho_t = \exp\left(\int_0^t b(X_s)\, dX_s - \frac{1}{2}\int_0^t b^2(X_s)\, d[X]_s\right), \quad t \in \mathbb{R}_+.$$

Indeed, under P, assuming $b(X) \in \Lambda(\mathcal{P}, X)$, the stochastic integral $\int_0^t b(X_s)\, dX_s$ defines a continuous local martingale, and by Theorem 6.2 and Corollary 5.9, $\{\rho_t, \mathcal{F}_t, t \in \mathbb{R}_+\}$ defined by (9.37) is a local martingale. Moreover, by (6.7) and the substitution theorem 2.12, it satisfies

$$(9.38) \qquad \rho_t = 1 + \int_0^t \rho_s b(X_s)\, dX_s \quad \text{for all } t \geq 0.$$

By (5.14) we have P-a.s.,

$$(9.39) \qquad [\rho, X]_t = \int_0^t \rho_s b(X_s)\, d[X]_s \quad \text{for all } t \in \mathbb{R}_+,$$

and hence

$$\int_0^t \rho_s^{-1} d[\rho, X]_s = \int_0^t b(X_s) d[X]_s \quad \text{for all } t \in \mathbb{R}_+.$$

We shall now apply Theorem 9.8 locally on each $\mathcal{F}_t$. For this we *suppose* $\{\rho_t, \mathcal{F}_t, t \in \mathbb{R}_+\}$ *is a martingale under* P (not just a local martingale — see the Remark below).

For each $t \in \mathbb{R}_+$, let P_t denote the restriction of P to $\mathcal{F}_t$ and let Q_t be the probability measure on $\mathcal{F}_t$ that is absolutely continuous with respect to P_t and whose Radon-Nikodym derivative is given by

$$(9.40) \qquad \frac{dQ_t}{dP_t} = \rho_t \quad \text{on } \mathcal{F}_t.$$

Since $\rho_t > 0$ P-a.s., Q_t is equivalent to P_t. Also, since $\{\rho_t, \mathcal{F}_t, t \in \mathbb{R}_+\}$ is a martingale under P, the Q_t's are consistent, i.e., if $0 \leq s < t$ and $F \in \mathcal{F}_s$, then

$$(9.41) \qquad \begin{aligned} Q_t(F) &= E^{P_t}[\rho_t 1_F] = E^P[\rho_t 1_F] \\ &= E^P[\rho_s 1_F] = E^{P_s}[\rho_s 1_F] = Q_s(F). \end{aligned}$$

Then by applying Theorem 9.8 with $\mathcal{F}_t, Q_t, \rho_t, \{\rho_{s \wedge t}, \mathcal{F}_{s \wedge t}, s \in I\!\!R_+\}, X_{\cdot \wedge t}$, $A_{\cdot \wedge t}$ in place of $\mathcal{F}, Q, \rho, \{\rho_t, \mathcal{F}_t, t \in I\!\!R_+\}, X, A$, respectively, we conclude that $\{X_{s \wedge t} - \int_0^{s \wedge t} b(X_u) \, d[X]_u, \mathcal{F}_{s \wedge t}, s \in I\!\!R_+\}$ is a continuous local martingale under Q_t with the same quadratic variation process as $X_{\cdot \wedge t}$ under P. Thus, under the change of measure from P_t to Q_t on $\mathcal{F}_t$, X restricted to the time interval $[0, t]$ is transformed from a continuous local martingale under P_t to a continuous local martingale plus a state-dependent drift of the form $\int_0^{\cdot \wedge t} b(X_u) \, d[X]_u$.

Remark. We could try to go one step further in the application of Theorem 9.8 and consider the case where $\{\rho_t, \mathcal{F}_t, t \in I\!\!R_+\}$ is just a local martingale. We could perform mutually absolutely continuous changes of probability over the σ-fields $\mathcal{F}_{t \wedge \tau_n}$, where $\{\tau_n\}$ is a localizing sequence for $\{\rho_t, t \in I\!\!R_+\}$. However, the associated measures $\{Q_{t \wedge \tau_n}, n \in I\!\!N\}$ will be extendible to a probability measure on each $\mathcal{F}_t$ only if $\{\rho_t, t \in I\!\!R_+\}$ is a martingale. The problem is that a process associated with the laws $Q_{t \wedge \tau_n}$ may run off to infinity at a finite random time. Such processes are said to *explode* and then a precise description would require an enlarged probability space containing paths that reach infinity in finite time and stay there forever after. We shall not pursue such a generalization here.

We now apply the above results to a Brownian motion.

Example. *(Brownian motion plus a drift.)* Suppose X is a one-dimensional Brownian motion on $(\Omega, \mathcal{F}, P)$ and $b : I\!\!R \to I\!\!R$ is Borel measurable.

First consider the case where b is bounded. Then $\{\rho_t, \mathcal{F}_t, t \in I\!\!R_+\}$ given by (9.26) is a continuous L^2-martingale under P, and we define Q_t and $B = \{B_s, s \in I\!\!R_+\}$ by (9.27) and (9.28) respectively. Then by the above application, on $(\Omega, \mathcal{F}_t, Q_t)$, $\{B_{s \wedge t}, \mathcal{F}_{s \wedge t}, s \in I\!\!R_+\}$ is a continuous local martingale with quadratic variation $[B_{\cdot \wedge t}]_s = s \wedge t$ Q_t-a.s. for all $s \in I\!\!R_+$. It then follows from Theorem 9.4 that under Q_t, $\{B_s, s \in [0, t]\}$ is a one-dimensional Brownian motion on the time interval $[0, t]$. Thus, the claims in the example at the beginning of this section are justified.

We now consider the case of an unbounded b. Assuming $X_0 \in L^2$, we will show that under certain conditions on b, $\{\rho_t, \mathcal{F}_t, t \in I\!\!R_+\}$ given by

(9.26) is a martingale on $(\Omega, \mathcal{F}, P)$, and then the above analysis applies. We suppose b is locally bounded, i.e., bounded on each compact set, and satisfies the growth condition that there is a constant $C > 0$ such that

$$(9.42) \qquad x \cdot b(x) \leq C(1 + x^2) \quad \text{for all } x \in \mathbb{R}.$$

Condition (9.42) is commonly referred to as a condition for non-explosion. If it is not satisfied, e.g., $b(x) = |x|^\delta$ for some $\delta > 1$, then a Brownian motion with such a drift may run off to infinity in a finite amount of time and its law cannot be obtained from that of a Brownian motion by absolutely continuous changes of probability on the σ-fields $\mathcal{F}_t$. For further discussion see Durrett [27, p. 240].

Assuming b is as described above and $X_0 \in L^2$, let $\tau_n = \inf\{t \geq 0 : X_t \notin (-n, n)\} \wedge n$ for each $n \in \mathbb{N}$. Since b is locally bounded, $b(X) \in \Lambda(\mathcal{P}, X)$ and for ρ_t given by (9.26), $\{\rho_t, \mathcal{F}_t, t \in \mathbb{R}_+\}$ is a continuous local L^2-martingale with $\{\tau_n\}$ as a localizing sequence. Then, by Proposition 1.8, to prove $\{\rho_t, \mathcal{F}_t, t \in \mathbb{R}_+\}$ is a martingale, it suffices to show that $\{\rho_{t \wedge \tau_n}, n \in \mathbb{N}\}$ is uniformly integrable for each $t \in \mathbb{R}_+$. The following proof of this uses a modification of an argument in [27, p. 240]. Note first that since $\rho_t \geq 0$, by Fatou's lemma and the martingale property of $\{\rho_{t \wedge \tau_n}, \mathcal{F}_t, t \in \mathbb{R}_+\}$ under P, we have

$$(9.43) \qquad E^P[\rho_t] = E^P[\lim_{n \to \infty} \rho_{t \wedge \tau_n}] \leq \liminf_{n \to \infty} E^P[\rho_{t \wedge \tau_n}] = 1.$$

Now consider Itô's formula (5.22) applied to the function $f(x_1, x_2, y) = x_1(1 + x_2^2)y$, and $M_t^1 = \rho_t$, $M_t^2 = X_t$ and $V_t^1 = \exp(-2(C+1)t)$. Then we have P-a.s. for all $t \in \mathbb{R}_+$,

$$\rho_t(1 + X_t^2)e^{-2(C+1)t} - (1 + X_0^2)$$

$$= \int_0^t (1 + X_s^2)e^{-2(C+1)s}\, d\rho_s + 2\int_0^t \rho_s X_s e^{-2(C+1)s}\, dX_s$$

$$(9.44) \qquad\qquad - 2(C+1)\int_0^t \rho_s(1 + X_s^2)e^{-2(C+1)s}\, ds$$

$$+ 2\int_0^t \rho_s e^{-2(C+1)s}\, ds + 2\int_0^t X_s e^{-2(C+1)s}\rho_s b(X_s)\, ds,$$

where we have used (9.39) to substitute for the integrator $[\rho, X]_s$ in the last integral. By (9.42), we have $X_s b(X_s) \le C(1 + X_s^2)$. Combining this with the non-negativity of the other terms in the last integral in (9.44), it follows that this integral is bounded above by $2C \int_0^t (1 + X_s^2) e^{-2(C+1)s} \rho_s \, ds$. Using this we conclude from (9.44) that

$$\rho_t(1 + X_t^2) e^{-2(C+1)t} - (1 + X_0^2)$$

$$\le \int_0^t (1 + X_s^2) e^{-2(C+1)s} \, d\rho_s + 2 \int_0^t \rho_s X_s e^{-2(C+1)s} \, dX_s.$$

Since this holds P-a.s. for all $t \in \mathbb{R}_+$, we can replace t with $t \wedge \tau_n$. Then, using (9.38) and the substitution theorem 2.12, we obtain P-a.s.,

(9.45)
$$\rho_{t \wedge \tau_n}(1 + X_{t \wedge \tau_n}^2) e^{-2(C+1)(t \wedge \tau_n)} - (1 + X_0^2)$$

$$\le \int_0^{t \wedge \tau_n} (1 + X_s^2) e^{-2(C+1)s} \rho_s b(X_s) dX_s + 2 \int_0^{t \wedge \tau_n} \rho_s X_s e^{-2(C+1)s} dX_s.$$

Now for fixed n, X and $b(X)$ are bounded on $(0, \tau_n]$, and $\rho_{\cdot \wedge \tau_n}$ defines an L^2-martingale. These properties, together with Doob's inequality (1.4) for $\rho_{\cdot \wedge \tau_n}$, can be used to verify that, when multiplied by $1_{(0,\tau_n]}$, the integrands in the stochastic integrals in (9.45) are in $\Lambda^2(\mathcal{P}, X)$. Consequently, as functions of t, these stochastic integrals define L^2-martingales and hence have zero expectations. Thus,

$$E^P \left[\rho_{t \wedge \tau_n}(1 + X_{t \wedge \tau_n}^2) e^{-2(C+1)(t \wedge \tau_n)} \right] \le E^P \left[1 + X_0^2 \right],$$

where the right member is finite by the assumption that $X_0 \in L^2$. Now for $n > t$, $X_{t \wedge \tau_n} \ge n$ on $\{\tau_n < t\}$, and then it follows from the above that

$$E^P \left[\rho_{\tau_n}(1 + n^2) e^{-2(C+1)t}; \tau_n < t \right] \le E^P \left[1 + X_0^2 \right].$$

Hence, for $n > t$,

(9.46)
$$E^P[\rho_{\tau_n}; \tau_n < t] \le e^{2(C+1)t} E^P[1 + X_0^2]/(1 + n^2).$$

Thus, for fixed $t \in \mathbb{R}_+$ and $K > 0$,

$$E^P \left[\rho_{t \wedge \tau_n}; \rho_{t \wedge \tau_n} > K \right] \le E^P \left[\rho_{\tau_n}; \tau_n < t \right] + E^P \left[\rho_t; \rho_t > K \right],$$

where by (9.46) the first term in the right member above can be made arbitrarily small for all n sufficiently large, and by (9.43) the second term can be made arbitrarily small by choosing K sufficiently large (independent of n). Recalling that ρ is non-negative, it follows that $\{\rho_{t \wedge T_n}, n \in I\!N\}$ is uniformly integrable and hence $\{\rho_t, \mathcal{F}_t, t \in I\!R_+\}$ is a martingale on $(\Omega, \mathcal{F}, P)$. Then, as in the case of b bounded, under the measure Q_t which is defined on $\mathcal{F}_t$ by (9.27) we have

$$X_s = B_s + \int_0^s b(X_u)du \quad \text{for } s \in [0, t],$$

where $\{B_s, s \in [0, t]\}$, is a Brownian motion on the time interval $[0, t]$.

It is natural to try to extend the consistent family of probability measures $\{Q_t, t \in I\!R_+\}$ to a probability measure Q on $\mathcal{F}_\infty$. However, this is possible only if $\{\rho_t, t \in I\!R_+\}$ is uniformly integrable. For if such a Q were to exist, then Q would be absolutely continuous with respect to P and the Radon-Nikodym derivative dQ/dP could be used to close the martingale $\{\rho_t, \mathcal{F}_t, t \in [0, \infty)\}$ at infinity. To see the absolute continuity of such an extension Q, suppose to the contrary that there is a P-null set $F \in \mathcal{F}_\infty$ with positive Q-probability. This F is in $\mathcal{F}_0$, which contains all P-null sets. Then by consistency and the fact that $\rho_0 = 1$, we have

$$0 < Q(F) = Q_0(F) = P_0(F) = P(F) = 0,$$

a contradiction. Then, assuming that Q is absolutely continuous with respect to P, and letting $\rho_\infty = dQ/dP$, we see that for $F \in \mathcal{F}_t$,

$$E^P[\rho_\infty 1_F] = Q(F) = Q_t(F) = E^{P_t}[\rho_t 1_F] = E^P[\rho_t 1_F],$$

and so $\rho_t = E^P[\rho_\infty \mid \mathcal{F}_t]$ for all $t \in I\!R_+$, which yields the uniform integrability of $\{\rho_t, t \in I\!R_+\}$.

The problem with extending the Q_t's to $\mathcal{F}_\infty \equiv \bigvee_{t \in I\!R_+} \mathcal{F}_t$ is that the σ-fields $\mathcal{F}_t$ contain all of the P-null sets. We can avoid this difficulty by restricting the probability measures Q_t to the raw (unaugmented) σ-fields generated by X on $[0, t]$, and then one can extend these restricted measures to a probability measure on the raw σ-field generated by X on $[0, \infty)$. This extension will not in general be absolutely continuous with respect to P. To facilitate the extension, we now assume a canonical representation for our Brownian motion X. There is no loss of generality in doing this, because we can always consider the probability measure induced on path space by X under P.

Suppose Ω is the set of all continuous functions $\omega : [0, \infty) \to I\!R$ and let $\mathcal{F}^o$ denote the raw σ-field $\sigma\{\omega(s) : 0 \le s < \infty\}$ and $\{\mathcal{F}_t^o, t \in I\!R_+\}$ denote the raw filtration defined by $\mathcal{F}_t^o = \{\omega(s) : 0 \le s \le t\}$ for each $t \in I\!R_+$. Let $X = \{X_t, t \in I\!R_+\}$ be the coordinate process on Ω defined by

$$X_t(\omega) = \omega(t) \quad \text{for all } t \in I\!R_+ \text{ and } \omega \in \Omega.$$

Suppose P is a probability measure on $(\Omega, \mathcal{F}^o)$ such that $X = \{X_t, \mathcal{F}_t^o, t \in I\!R_+\}$ is a Brownian motion martingale under P. Further assume $X_0 \in L^2$ if b is not bounded. Let $\mathcal{F}$ denote the completion of $\mathcal{F}^o$ under P and let $\mathcal{F}_t$ denote the augmentation of $\mathcal{F}_t^o$ by all of the P-null sets in $\mathcal{F}$. Then $(\Omega, \mathcal{F}, P)$ is complete and the filtration $\{\mathcal{F}_t, t \in I\!R_+\}$ is standard [12, p. 61] and so all of the above analysis applies to it. The key to the extension is the following.

Proposition 9.9. *Any consistent family $\{\nu_t, t \in I\!R_+\}$ of probability measures defined on $\{\mathcal{F}_t^o, t \in I\!R_+\}$ can be uniquely extended to a probability measure on $\mathcal{F}^o$.*

Proof. The uniqueness is clear, since the $\mathcal{F}_t^o$'s generate $\mathcal{F}^o$. For the existence, note that by Kolmogorov's extension theorem there is a unique probability measure on the product space $I\!R^{[0,\infty)}$ whose finite dimensional distributions are consistent with the family $\{\nu_t, t \in I\!R_+\}$. Moreover, there is a continuous version of a process associated with this measure, because there is a continuous version on each $[0, t]$, by the existence of the ν_t's. The probability measure induced on $(\Omega, \mathcal{F}^0)$ by such a continuous version is the required extension of the ν_t's. ∎

Let the Q_t's be defined on the $\mathcal{F}_t$'s as above under the assumption that b is locally bounded and satisfies (9.42). Then by Proposition 9.9, there is a unique extension of the restrictions of the Q_t's on the $\mathcal{F}_t^o$'s, to a probability measure Q on $\mathcal{F}^o$. Since $\{B_s = X_s - \int_0^s b(X_u)\, du, s \in [0, t]\}$ is

$\mathcal{F}_t^o$-measurable and has the law of a Brownian motion under Q_t, for each t. It follows that $\{B_s, s \in \mathbb{R}_+\}$ is a Brownian motion on $(\Omega, \mathcal{F}^o)$ under Q. Thus, on $(\Omega, \mathcal{F}^o, Q)$,

$$X_t = B_t + \int_0^t b(X_s)\,ds \quad \text{for all } t \in \mathbb{R}_+,$$

where B is a Brownian motion. In other words, under Q, X is a solution of the stochastic differential equation

$$dX_t = dB_t + b(X_t)\,dt,$$

where B is a Brownian motion. We say that X is a *weak* solution of this stochastic differential equation because we are free to give both X and B in specifying it. If X were produced from a given Brownian motion B, then it would be called a *strong* solution.

Remark. If necessary, the probability space $(\Omega, \mathcal{F}^o, Q)$ can be completed and then the filtration $\{\mathcal{F}_{t+}^o\}$ augmented with the Q-null sets to ensure a complete probability space and standard filtration under Q. This should be done for instance before applying the Itô formula under Q, since this formula was developed under the "usual conditions".

Two simple examples of the application of the above results are given below. Here X is a one-dimensional Brownian motion under P on the canonical space $(\Omega, \mathcal{F}^o)$ and Q is defined immediately above.

Example 1. (*Brownian motion with constant drift.*) If $b(\cdot) \equiv \mu \in \mathbb{R}\backslash\{0\}$, then

$$\rho_t = \exp\left(\mu(X_t - X_0) - \frac{1}{2}\mu^2 t\right)$$

defines an L^2-martingale under P, and $X_t - \mu t$ is a Brownian motion on $(\Omega, \mathcal{F}^o, Q)$. The probability measures Q and P are not absolutely continuous on $\mathcal{F}^o$ because

$$P(\lim_{t\to\infty} X_t/t = 0) = 1 \quad \text{and} \quad Q(\lim_{t\to\infty} X_t/t = \mu \neq 0) = 1.$$

Example 2. (*Ornstein-Uhlenbeck process.*) Assume $b(x) = \alpha x$ for some

$\alpha \in I\!\!R \backslash \{0\}$, and $X_0 \in L^2$. Then b is locally bounded, it satisfies (9.42),

$$\rho_t = \exp\left(\alpha \int_0^t X_s \, dX_s - \frac{\alpha^2}{2} \int_0^t X_s^2 \, ds\right)$$

defines a martingale under P, and on $(\Omega, \mathcal{F}^o, Q)$, $X_t = B_t + \alpha \int_0^t X_s \, ds$, where B is a Brownian motion. Thus, under Q, X is a weak solution of the stochastic differential equation:

$$dX_t = dB_t + \alpha X_t \, dt,$$

where B is a Brownian motion. The reader should compare this with the strong solution exhibited in Section 5.2.

In a similar manner to the above example for one-dimensional Brownian motion, a change of measure transformation can be derived for d-dimensional Brownian motion and a locally bounded drift $b : I\!\!R^d \to I\!\!R^d$ that satisfies a multi-dimensional analogue of (9.42). The characterization of Brownian motion required for this is a version of Exercise 2 in Chapter 6 for a finite interval. Specifically, a d-tuple $(M^1, \ldots, M^d)$ of continuous local martingales, with mutual variation given by $[M^i, M^j]_s = \delta_{ij} s$ for $s \in [0, t]$, has the law of a d-dimensional Brownian motion on the time interval $[0, t]$. For reference, we state one form of the d-dimensional change of measure result below. This also serves to summarize some of the preceding results for the case $d = 1$.

Theorem 9.10. *Suppose $X = \{X(t, \omega) = \omega(t), t \in I\!\!R_+, \omega \in \Omega\}$ is a d-dimensional Brownian motion on the probability space $(\Omega, \mathcal{F}^o, P)$, where $\Omega = \{\omega : [0, \infty) \to I\!\!R^d, \omega$ is continuous$\}$ and $\mathcal{F}^o = \sigma\{\omega(s) : 0 \le s < \infty\}$. Let $b : I\!\!R^d \to I\!\!R^d$ be a locally bounded Borel measurable function satisfying*

$$x \cdot b(x) \le C(1 + |x|^2) \quad \text{for all } x \in I\!\!R^d,$$

for some $C > 0$. If b is not bounded, also assume that $X_0 \in L^2$. Then there is a probability measure Q on $(\Omega, \mathcal{F}^o)$, equivalent to P on each $\mathcal{F}_t^o \equiv \sigma\{\omega(s) : 0 \le s \le t\}$, such that

$$X_t = B_t + \int_0^t b(X_s) \, ds \quad \text{for all } t \in I\!\!R_+,$$

where B is a d-dimensional Brownian motion under Q. The relationship between P and Q is determined by

$$\frac{dQ}{dP} = \exp\left(\int_0^t b(X_s)dX_s - \frac{1}{2}\int_0^t |b|^2(X_s)\, ds \right) \quad \text{on } \mathcal{F}_t^o \text{ for each } t \in \mathbb{R}_+,$$

where the stochastic integral

$$\int_0^t b(X_s)dX_s \equiv \sum_{i=1}^d \int_0^t b_i(X_s)dX_s^i$$

is defined on the completion of $(\Omega, \mathcal{F}^o, P)$.

9.5 Exercises

1. Suppose $\{\tau_t, t \in [0, \infty)\}$ is an increasing family of optional times relative to the right continuous filtration $\{\mathcal{F}_t, t \geq 0\}$ and suppose $t \to \tau_t$ is right continuous. Show that $\{\mathcal{F}_{\tau_t}, t \geq 0\}$ is right continuous.

2. Show that the t_k defined in the proof of Theorem 9.3 are optional relative to $\{\mathcal{F}_{\tau_t}, t \in [0, \infty)\}$.

3. Show that $\{\rho_t, t \geq 0\}$ defined by (9.26) is not uniformly integrable when $b(x) \equiv b$, a non-zero constant. (*Hint:* Use the martingale convergence theorem.)

4. A solution of the following stochastic differential equation (s.d.e.) is called a Brownian motion with *bang-bang* drift:

$$dX_t = -\,\text{sgn}\,(X_t)dt + dB_t.$$

Here B is a one-dimensional Brownian motion. One can think of X as a Brownian motion controlled by a state-dependent drift that has magnitude one and is always directed towards the origin. Use a change of measure transformation to exhibit a weak solution of this s.d.e. Then use Tanaka's formula and the result of Exercise 1 in Chapter 8 to deduce the distribution of the solution X_t for any $t > 0$, assuming $X(0) = 0$ a.s.

Remark: It will be seen in Chapter 10 that any solution of this s.d.e. is unique in law, given the initial distribution of X.

5. Prove the change of measure result for d-dimensional Brownian motion which is described at the end of the first example in Section 9.4. If you are ambitious, prove Theorem 9.10, including the extension of Exercise 2 of Chapter 6 described immediately before the theorem.

10

STOCHASTIC DIFFERENTIAL
EQUATIONS

10.1 Introduction

In this chapter, we consider *stochastic differential equations* (SDE's) of the form

$$(10.1) \qquad dX(t) = \sigma(X(t))dB(t) + b(X(t))dt,$$

or equivalently in coordinate form

$$(10.2) \quad dX_i(t) = \sum_{j=1}^{r} \sigma_{ij}(X(t))dB_j(t) + b_i(X(t))dt, \quad \text{for } i = 1, \ldots, d,$$

where $B = (B_1, \ldots, B_r)$ is an r-dimensional Brownian motion ($r \geq 1$) starting from the origin, and $\sigma : \mathbb{R}^d \to \mathbb{R}^d \otimes \mathbb{R}^r$ and $b : \mathbb{R}^d \to \mathbb{R}^d$ are Borel measurable functions. Here $\mathbb{R}^d \otimes \mathbb{R}^r$, $d \geq 1$, $r \geq 1$, denotes the space of $d \times r$ real-valued matrices with the norm

$$(10.3) \qquad \|A\| = \left(\sum_{i=1}^{d} \sum_{j=1}^{r} A_{ij}^2 \right)^{\frac{1}{2}},$$

for $A \in \mathbb{R}^d \otimes \mathbb{R}^r$.

Notation. In this chapter we shall no longer denote a multidimensional Brownian motion in bold, and we shall use subscripts rather than superscripts to denote the components of a multidimensional process.

Stochastic differential equations of the form (10.1) are used to describe the dynamics of a wide variety of random phenomena in the physical, biological, engineering and social sciences. Under suitable conditions on the coefficients, solutions of (10.1) with initial conditions running over the points in the state space generate a continuous strong Markov process, otherwise known as a *diffusion process*. In this case, assuming b and σ are bounded, say, so that the following expectations are finite, we have

$$E[X_i(t+h) - X_i(t)|X(s) : 0 \le s \le t] = b_i(X(t))h + o(h)$$

and

$$E[(X_i(t+h)-X_i(t))(X_j(t+h)-X_j(t)) \mid X(s) : 0 \le s \le t] = a_{ij}(X(t))h+o(h)$$

as $h \to 0$, for $i, j = 1, \ldots, d$, where $a = \sigma\sigma'$ and $'$ denotes transpose. The coefficient b is referred to as the *drift vector*, and a is called the *diffusion matrix* or infinitesimal covariance matrix. The latter determines the mutual variation process of X: $d\langle X_i, X_j \rangle_t = a_{ij}(X_t)dt$.

Remark. Here we have used the term diffusion process in its widest sense to mean a continuous strong Markov process. The reader should be warned however that there is no commonly agreed convention as to the applicability of this term. For instance, some authors use it only for diffusions of the type considered in this chapter, namely those continuous strong Markov processes arising from solutions of SDE's of the form (10.1). This would exclude important continuous strong Markov processes such as those that behave like a solution of (10.1) in the interior of some Euclidean domain, but are confined to the closure of the domain by absorption or reflection at the boundary. The reflected Brownian motions studied in Chapter 8 are examples of the latter.

The rigorous interpretation of (10.1) is as a stochastic integral equation of the form

(10.4) $X(t) = X(0) + \int_0^t \sigma(X(s))dB(s) + \int_0^t b(X(s))ds$,   for all $t \geq 0$.

Notation. In accordance with (10.2), the vector stochastic integral above is defined by

(10.5) $\left(\int_0^t \sigma(X(s))dB(s) \right)_i = \sum_{j=1}^r \int_0^t \sigma_{ij}(X(s))dB_j(s)$, for $i = 1, \ldots, d$.

Of course, the process X, the coefficients σ and b, and the initial value of X, will need to satisfy certain conditions in order that the integrals in (10.4) are well defined.

Our aim here is not to give an exhaustive treatment of stochastic differential equations, but rather to give an introduction to the subject and to give some examples. Consequently, we shall largely confine our discussion to the classic case first considered by Itô, in which the coefficients satisfy a Lipschitz condition. If the Lipschitz condition is not uniform, then a growth condition on the coefficients at infinity will also be assumed. In Section 10.2 we prove existence and uniqueness of solutions of (10.1) under such conditions using a stochastic analogue of the Picard iteration scheme for solving ordinary differential equations. The solutions obtained in this manner are adapted to the filtration generated by the Brownian motion B and the initial random variable $X(0)$. They are *strong* solutions of the SDE (10.1) and they generate a diffusion process. The latter is proved in Section 10.3. A discussion of strong versus weak solutions of (10.1) and the associated notions of uniqueness is given in Section 10.4. The most important difference between the two notions is that a strong solution is defined on a given probability space with a given Brownian motion B and independent initial random variable $X(0)$, and it is adapted to the filtration generated by B and $X(0)$; whereas a weak solution may be defined on any probability space, where the Brownian motion B is defined as part of the solution. We conclude the chapter with Section 10.5 which contains some examples.

Terminology. In the sequel, we shall use the following terminology. By a filtered probability space we shall mean a complete probability space $(\Omega, \mathcal{F}, P)$ together with a *standard* filtration $\{\mathcal{F}_t\}$ (see Section 1.4 for the definition of a standard filtration). For $n \geq 1$, an n-dimensional process will be said to have a certain property such as boundedness, integrability, being a martingale, etc., if and only if each of its components has the property. A Brownian motion martingale on a filtered probability space $(\Omega, \mathcal{F}, \{\mathcal{F}_t\}, P)$ is a Brownian motion that is a martingale relative to the standard filtration $\{\mathcal{F}_t\}$. All of the stochastic integration results developed in previous chapters for a Brownian motion relative to its natural filtration (cf. §1.8) also hold for a Brownian motion martingale. In addition, by the proof of Theorem 6.1, the increments of such a Brownian motion martingale are independent of any $\mathcal{F}_0$-measurable random variable.

10.2 Existence and Uniqueness for Lipschitz Coefficients

We consider first the case where the coefficients $\sigma : \mathbb{R}^d \to \mathbb{R}^d \otimes \mathbb{R}^r$ and $b : \mathbb{R}^d \to \mathbb{R}^d$ are *bounded* and satisfy the uniform Lipschitz conditions:

$$(10.6) \qquad \|\sigma(x) - \sigma(y)\| \leq K|x - y|,$$

$$(10.7) \qquad |b(x) - b(y)| \leq K|x - y|,$$

for some constant $K > 0$ and all $x, y \in \mathbb{R}^d$. Here $|\cdot|$ denotes the usual Euclidean norm and $\|\cdot\|$ is the matrix norm defined in (10.3).

Let B be an r-dimensional Brownian motion martingale on some filtered probability space $(\Omega, \mathcal{F}, \{\mathcal{F}_t\}, P)$, and suppose that B starts from the origin. Let $X(0)$ be an $\mathcal{F}_0$-measurable random variable taking values in $\mathbb{R}^d$. Note that since B is a martingale relative to $\{\mathcal{F}_t\}$ then $B(\cdot) = B(\cdot) - B(0)$ is independent of $X(0)$.

Remark. One way to realize this setup is to take a Brownian motion B starting from the origin, and an independent random variable $X(0)$, both defined on some probability space $(\Omega, \mathcal{G}, P)$. One can then choose $\mathcal{F}$ to be the P-completion of the σ-field generated by $X(0)$ and B, and $\mathcal{F}_t$ to be the σ-field generated by $X(0)$ and $\{B(s) : 0 \leq s \leq t\}$, augmented by the P-null sets in $\mathcal{F}$. For deterministic initial conditions: $X(0) \equiv x \in I\!\!R^d$, one can simply take the completed σ-field and augmented filtration generated by B alone.

We shall now prove that there is a unique solution of the stochastic integral equation (10.4) when σ and b are bounded and satisfy (10.6)–(10.7).

Denote the right member of (10.4) by $\tilde{X}(t)$. We first show that $\tilde{X}$ is well defined for any $(\mathcal{B} \times \mathcal{F})$-measurable adapted process X. For such an X, each component of X is $\mathcal{P}^*$-measurable and hence so is each component of $\sigma(X)$, since σ is continuous. This, together with the boundedness of σ, implies that each of the stochastic integrals in the right member of (10.5) is well defined as a continuous L^2-martingale. Moreover, by Cauchy's inequality and the L^2-isometry for stochastic integrals, we have for each $t \in I\!\!R_+$,

$$
\begin{aligned}
E\left[\left|\int_0^t \sigma(X(s))dB(s)\right|^2\right] &= E\left[\sum_{i=1}^d \left(\int_0^t \sigma(X(s))dB(s)\right)_i^2\right] \\
&= E\left[\sum_{i=1}^d \left(\sum_{j=1}^r \int_0^t \sigma_{ij}(X(s))dB_j(s)\right)^2\right] \\
&\leq \sum_{i=1}^d r \sum_{j=1}^r E\left[\left(\int_0^t \sigma_{ij}(X(s))dB_j(s)\right)^2\right] \\
&= r\sum_{i=1}^d \sum_{j=1}^r E\left[\int_0^t \sigma_{ij}^2(X(s))ds\right] \\
&= rE\left[\int_0^t \|\sigma(X(s))\|^2\,ds\right] < \infty.
\end{aligned}
$$

(10.8)

Since X is $(\mathcal{B} \times \mathcal{F})$-measurable and b is continuous, $b(X)$ is $(\mathcal{B} \times \mathcal{F})$-measurable. Combining this with the boundedness of b, we see that $\{\int_0^t b(X(s))ds,\ t \geq 0\}$ is defined path-by-path as a continuous d-dimensional process on $(\Omega, \mathcal{F}, P)$. We do not know a priori that the process is adapted. The following argument shows this.

Since X is $\mathcal{P}^*$-measurable, there is a $\mathcal{P}$-measurable random variable Y such that $X = Y$, $(\lambda \times P)$-a.e. Now any $\mathcal{P}$-measurable random variable is progressively measurable (cf. Section 3.3), and so $Y|_{[0,t]\times\Omega}$ is $(\mathcal{B}_t \times \mathcal{F}_t)$-measurable, where $\mathcal{B}_t$ is the Borel σ-field on $[0,t]$. Thus $b(Y)|_{[0,t]\times\Omega}$ is also $(\mathcal{B}_t \times \mathcal{F}_t)$-measurable, since b is continuous. Then by the Fubini-Tonelli theorem, for each $t \in [0,\infty)$, $\int_0^t b(Y(s))ds$ defines an $\mathcal{F}_t$-measurable random variable. Now we have

$$E\left[\left|\int_0^t b(X(s))ds - \int_0^t b(Y(s))ds\right|\right]$$
$$\leq E\left[\int_0^t |b(X(s)) - b(Y(s))|\,ds\right]$$
$$\leq KE\left[\int_0^t |X(s) - Y(s)|\,ds\right]$$
$$= 0,$$

where we have used (10.7) for the second inequality and the fact that $X = Y$ ($\lambda \times P$)-a.e. for the last equality. Thus $\int_0^t b(X(s))ds = \int_0^t b(Y(s))ds$ P-a.s. Since $\mathcal{F}_t$ contains all of the P-null sets, it follows that $\int_0^t b(X(s))ds$ is $\mathcal{F}_t$-measurable. Thus $\{\int_0^t b(X(s))ds,\ t \geq 0\}$ is a continuous adapted process that is locally of bounded variation.

In summary, we have shown that if X is a $(\mathcal{B} \times \mathcal{F})$-measurable adapted process, then $\tilde{X}$ is well defined as a continuous adapted semimartingale process (see Exercise 3 of Chapter 4 for the definition of a semimartingale). If we remove the boundedness assumption on σ and b and do not assume (10.6) and (10.7), but assume rather that X is continuous and σ and b are continuous, then $\sigma(X)$ and $b(X)$ are continuous and the stochastic integral $\int_0^t \sigma(X(s))dB(s)$ defines a continuous *local* martingale (cf. Theorem 2.11), and the path-by-path integral $\int_0^t b(X(s))ds$ defines a continuous adapted process that is locally of bounded variation. This situation will be encountered later in this section where σ and b are only locally Lipschitz continuous.

The following lemma is fundamental to the proofs of existence and uniqueness of a solution to (10.4).

Lemma 10.1. *For each $T > 0$ there exists a constant C_T such that for any Y and Z that are $(\mathcal{B} \times \mathcal{F})$-measurable adapted d-dimensional processes* satisfying $Y_0 - Z_0 \in L^2$, we have for each $t \in [0, T]$,

$$(10.9) \quad E\left[\sup_{0 \leq s \leq t} |\tilde{Y}_s - \tilde{Z}_s|^2\right] \leq 3E\left[|Y_0 - Z_0|^2\right] + C_T E\left[\int_0^t |Y_s - Z_s|^2 \, ds\right],$$

where for each $t \geq 0$, $\tilde{Y}(t)$, respectively $\tilde{Z}(t)$, denotes the right member of (10.4) with Y, respectively Z, in place of X there.

Proof. Without loss of generality we may suppose the last written expectation is finite. Then by the definition of $\tilde{Y}$ and $\tilde{Z}$, we have

$$(10.10) \quad |\tilde{Y}_t - \tilde{Z}_t|^2 \leq 3\left\{|Y_0 - Z_0|^2 + \left|\int_0^t (\sigma(Y_s) - \sigma(Z_s))dB_s\right|^2 \right.$$
$$\left. + \left|\int_0^t (b(Y_s) - b(Z_s))ds\right|^2\right\}.$$

In a similar manner to that leading to (10.8), since σ is bounded, the stochastic integral with respect to dB_s above defines a continuous L^2-martingale with

$$(10.11) \quad E\left[\left|\int_0^t (\sigma(Y_s) - \sigma(Z_s))dB_s\right|^2\right] \leq rE\left[\int_0^t \|\sigma(Y_s) - \sigma(Z_s)\|^2 \, ds\right].$$

By (10.6), the right member of (10.11) is dominated by

$$rK^2 E\left[\int_0^t |Y_s - Z_s|^2 ds\right].$$

Combining this with Doob's inequality (1.4) (which also applies to d-dimensional martingales), we obtain

$$(10.12) \quad E\left[\sup_{0 \leq s \leq t}\left|\int_0^s (\sigma(Y_u) - \sigma(Z_u))dB_u\right|^2\right] \leq 4rK^2 E\left[\int_0^t |Y_s - Z_s|^2 ds\right].$$

Now by the Cauchy-Schwarz inequality and (10.7) we have

$$
(10.13) \qquad
\begin{aligned}
E\left[\left|\int_0^t (b(Y_s) - b(Z_s))ds\right|^2\right] &\le E\left[t\int_0^t |b(Y_s) - b(Z_s)|^2 ds\right] \\
&\le K^2 t E\left[\int_0^t |Y_s - Z_s|^2 ds\right].
\end{aligned}
$$

Thus (10.9) follows with $C_T = 3K^2(4r + T)$. ∎

Remark. This lemma also holds if σ and b are not bounded, but still satisfy (10.6)–(10.7), and Y and Z are continuous. This can be proved using a localizing sequence together with Fatou's lemma (see Exercise 1).

Next we need an analytic lemma known as *Gronwall's inequality* (see Exercise 2).

Lemma 10.2. *Suppose f and g are Lebesgue integrable in $[0, T]$ for some $T \in (0, \infty)$ and there is a constant $C > 0$ such that*

$$
f(t) \le g(t) + C\int_0^t f(s)ds \quad \text{for all } t \in [0, T].
$$

Then

$$
f(t) \le g(t) + C\int_0^t e^{C(t-s)}g(s)ds \text{ for all } t \in [0, T].
$$

In particular, if there is a constant A such that $g(t) = A$ for all $t \in [0, T]$, then

$$
(10.14) \qquad f(t) \le Ae^{Ct} \text{ for all } t \in [0, T].
$$

Theorem 10.3. *There is at most one $(\mathcal{B} \times \mathcal{F})$-measurable adapted solution X of (10.4).*

Remark. As we discussed above, such a solution is automatically continuous because the right member of (10.4) is continuous for any $(\mathcal{B} \times \mathcal{F})$-measurable adapted X.

Proof. If Y and Z are two solutions of (10.4) then $Y_0 = X_0$, $Z_0 = X_0$ and $Y_t = \tilde{Y}_t$, $Z_t = \tilde{Z}_t$ for all t. Then for any $T > 0$, by (10.9) and Fubini's theorem, we have

$$E[|Y_t - Z_t|^2] \leq C_T \int_0^t E[|Y_s - Z_s|^2]ds \quad \text{for all } t \in [0, T].$$

Applying Lemma 10.2 with $f(t) = E[|Y_t - Z_t|^2]$ and $g(t) \equiv 0$ yields $f(t) = 0$ for all $t \in [0, T]$. Thus $P(Y_t = Z_t) = 1$ for all $t \in [0, T]$. Since $T > 0$ was arbitrary, this holds for all $t \in [0, \infty)$. Since Y and Z are continuous processes, it follows that they are indistinguishable. Alternatively, one can use

$$E\left[\sup_{0 \leq s \leq t} |Y_s - Z_s|^2 \right] \leq C_T \int_0^t E\left[\sup_{0 \leq u \leq s} |Y_u - Z_u|^2 \right] ds,$$

in conjunction with Lemma 10.2 to reach the same conclusion. ∎

Theorem 10.4. *There exists a continuous adapted solution X of (10.4). Moreover, its law is uniquely determined by that of $X(0)$ and B.*

Proof. We define a sequence of successive approximations to a solution, much as in Picard's method of solving deterministic differential equations. In fact, the estimation below is practically the same as in the classical case, the stochastic integral terms being estimated by the inequality (10.8).

Define $X_t^0 = X(0)$ for all $t \in [0, \infty)$, and define inductively for $n \geq 1$: $X^n = \tilde{X}^{n-1}$. For each $t \in (0, \infty)$,

(10.15)
$$X_t^1 = X_0 + \int_0^t \sigma(X_0)dB_s + \int_0^t b(X_0)ds$$

$$= X_0 + \sigma(X_0)B_t + b(X_0)t.$$

Here we have denoted $X(0)$ by X_0 and we have used the fact that $B(0) = 0$.

Let us note that each X^n is continuous and adapted. This follows from the paragraph preceding Lemma 10.1 and induction on n. Note in particular that if we take $\mathcal{F}_t$ to be the standard filtration generated by $X(0)$ and B, then X^n is adapted to this filtration. Moreover, by induction (writing the integrals in $\tilde{X}^{n-1}$ as limits of simple integrals), the law of X^n is uniquely determined by the law of $X(0)$ and the law of B, where $X(0)$ and B are independent.

Now from (10.15) we have

$$E\left[\sup_{0\le s\le t} |X^1_s - X^0_s|^2\right]$$

$$\le 2C_\sigma E\left[\sup_{0\le s\le t} |B_s|^2\right] + 2E[|b(X_0)|^2]t^2$$

(10.16)

$$\le 8C_\sigma E[|B_t|^2] + 2E[|b(X_0)|^2]t^2$$

$$\le C(t + t^2),$$

where C_σ is a constant depending only on the bound for σ and C is a constant depending on the bounds for σ and b. If σ and b are not bounded, the above inequality still holds if X_0 is a fixed constant, and the rest of the proof can be carried out without the boundedness of σ and b, using the Remark following Lemma 10.1.

Applying Lemma 10.1 with $Y = X^n$ and $Z = X^{n-1}$, we obtain for each $T \in [0, \infty)$:

$$(10.17) \qquad E\left[\sup_{0\le s\le t} |X^{n+1}_s - X^n_s|^2\right] \le C_T E\left[\int_0^t |X^n_s - X^{n-1}_s|^2 ds\right]$$

for all $t \in [0, T]$. Then, letting $D_{n+1}(t)$ denote the left member of (10.17), we have for $n \ge 1$,

$$(10.18) \qquad D_{n+1}(t) \le C_T \int_0^t D_n(s)ds \quad \text{for all } t \in [0, T],$$

and $D_1(t)$ is estimated in (10.16). By iteration we obtain

$$D_{n+1}(t) \leq C_T^n \int_0^t \int_0^{s_1} \cdots \int_0^{s_{n-1}} D_1(s)ds\,ds_{n-1}\ldots ds_1$$

(10.19)
$$\leq C_T^n C \int_0^t \int_0^{s_1} \cdots \int_0^{s_{n-1}} (s+s^2)ds\,ds_{n-1}\ldots ds_1 .$$

$$= CC_T^n \left(\frac{t^{n+1}}{(n+1)!} + \frac{2t^{n+2}}{(n+2)!} \right)$$

By the definition of $D_{n+1}(t)$ and Chebyshev's inequality we have for each $T > 0$,

(10.20)
$$P\left(\sup_{0 \leq s \leq T} |X_s^{n+1} - X_s^n| \geq \frac{1}{2^n} \right) \leq 2^{2n} D_{n+1}(T).$$

The estimate (10.19) shows that for fixed T, the right member above is the general term of a convergent series in n. Hence, by the Borel-Cantelli lemma,

(10.21)
$$P\left(\sup_{0 \leq s \leq T} |X_s^{n+1} - X_s^n| \geq \frac{1}{2^n} \text{ for infinitely many } n \right) = 0.$$

This implies that a.s., X_s^n converges to a limit X_s, uniformly for $s \in [0,T]$. Since $T > 0$ was arbitrary, it follows that a.s., X^n converges uniformly on any bounded time interval to a limit process X. This process inherits the continuity and adaptedness properties of the X^n. Since $X = \lim_{n \to \infty} X^n$, where the law of X^n is determined by the laws of $X(0)$ and B, it follows that the same is true for X.

To verify that X is a solution of (10.4), consider positive integers $m < n$. Then, since

$$\sup_{0 \leq s \leq T} |X_s^n - X_s^m| \leq \sum_{k=m+1}^n \sup_{0 \leq s \leq T} |X_s^k - X_s^{k-1}|,$$

by Minkowski's inequality we have

$$(10.22) \quad \left(E\left[\sup_{0 \le s \le T} |X_s^n - X_s^m|^2 \right] \right)^{\frac{1}{2}} \le \sum_{k=m+1}^{n} \left(E\left[\sup_{0 \le s \le T} |X_s^k - X_s^{k-1}|^2 \right] \right)^{\frac{1}{2}}$$

$$= \sum_{k=m+1}^{n} D_k^{\frac{1}{2}}(T).$$

Letting $n \to \infty$, by Fatou's lemma we obtain

$$(10.23) \quad E\left[\sup_{0 \le s \le T} |X_s - X_s^m|^2 \right] \le \left(\sum_{k=m+1}^{\infty} D_k^{\frac{1}{2}}(T) \right)^2 \equiv \varepsilon_m,$$

where $\varepsilon_m \to 0$ as $m \to \infty$. Applying Lemma 10.1 again with $Y = X$ and $Z = X^m$, and using (10.23), we obtain

$$E\left[\sup_{0 \le s \le T} |\tilde{X}_s - X_s^{m+1}|^2 \right] \le C_T \int_0^T \varepsilon_m \, ds = C_T T \varepsilon_m.$$

Letting $m \to \infty$ and using Fatou's lemma we conclude that

$$E\left[\sup_{0 \le s \le T} |\tilde{X}_s - X_s|^2 \right] = 0.$$

Thus $\tilde{X}$ and X are indistinguishable on $[0, T]$ for each $T > 0$ and hence on $[0, \infty)$. Recalling the definition of $\tilde{X}$, we realize that $X = \tilde{X}$ means X is a solution of (10.4). ∎

Combining Theorems 10.3, 10.4 and Lemmas 10.1, 10.2, we obtain the following.

Theorem 10.5. *Suppose σ and b are bounded and satisfy the uniform Lipschitz conditions (10.6)–(10.7). Let B be a Brownian motion martingale on a filtered probability space $(\Omega, \mathcal{F}, \{\mathcal{F}_t\}, P)$ such that $B(0) = 0$, P-a.s., and let $X(0)$ be an an $\mathcal{F}_0$-measurable random vector. Then there is a unique $(\mathcal{B} \times \mathcal{F})$-measurable, adapted solution X of (10.4). This solution is continuous, and its law is uniquely determined by σ, b and the laws of $X(0)$ and B. Moreover, if Y is a $(\mathcal{B} \times \mathcal{F})$-measurable adapted solution of (10.4)* *with the $\mathcal{F}_0$-measurable initial random vector $Y(0)$ in place of $X(0)$, then assuming $X(0) - Y(0) \in L^2$, we have for all $T \geq 0$,*

$$(10.24) \qquad E\left[\sup_{0 \leq s \leq T} |X(s) - Y(s)|^2\right] \leq 3E\left[|X(0) - Y(0)|^2\right] e^{C_T T}.$$

We now turn to the case where σ and b are locally Lipschitz continuous and satisfy a growth condition at infinity.

Suppose $\sigma : I\!\!R^d \to I\!\!R^d \otimes I\!\!R^r$ and $b : I\!\!R^d \to I\!\!R^d$ satisfy the conditions that for each $R > 0$ there is a constant $K_R > 0$ such that for all $x, y \in I\!\!R^d$ satisfying $|x|, |y| \leq R$,

$$(10.25) \qquad \|\sigma(x) - \sigma(y)\| \leq K_R|x - y|,$$

$$(10.26) \qquad |b(x) - b(y)| \leq K_R|x - y|,$$

and there is $K > 0$ such that for all $x \in I\!\!R^d$,

$$(10.27) \qquad \|\sigma(x)\|^2 \leq K(1 + |x|^2),$$

$$(10.28) \qquad x \cdot b(x) \leq K(1 + |x|^2).$$

It follows from (10.25)–(10.26) that σ and b are continuous. Note also that if

$$|b(x)|^2 \leq C(1 + |x|^2),$$

then

$$x \cdot b(x) \leq \frac{1}{2}(|x|^2 + |b(x)|^2) \leq \frac{C+1}{2}(1 + |x|^2).$$

Thus (10.28) is weaker than the analog of (10.27) for b. Note also that if σ and b satisfy the uniform Lipschitz conditions (10.6)–(10.7), then (10.25)–(10.28) hold.

Theorem 10.6. *Suppose σ, b satisfy (10.25)–(10.28), and B and $X(0)$ are as before. Then there is a unique continuous adapted solution X of (10.4). Moreover, the law of X is uniquely determined by the laws of $X(0)$ and B.*

Remark. We assume X is continuous in Theorem 10.6 to ensure that the integrals in (10.4) are well defined.

The following lemma is key to the proof of Theorem 10.6.

Lemma 10.7. *Fix $n \in \mathbb{N}$. Let $\sigma_Y, \sigma_Z : \mathbb{R}^d \to \mathbb{R}^d \otimes \mathbb{R}^r$ and $b_Y, b_Z : \mathbb{R}^d \to \mathbb{R}^d$ be continuous functions such that $\sigma_Y = \sigma_Z = \sigma$, $b_Y = b_Z = b$ on $\{x \in \mathbb{R}^d : |x| \leq n\}$. Suppose Y and Z are two continuous adapted processes such that for all $t \in [0, \infty)$:*

$$(10.29) \qquad Y(t) = Y(0) + \int_0^t \sigma_Y(Y(s))dB(s) + \int_0^t b_Y(Y(s))ds$$

$$(10.30) \qquad Z(t) = Z(0) + \int_0^t \sigma_Z(Z(s))dB(s) + \int_0^t b_Z(Z(s))ds.$$

Define $\tau_Y = \inf\{s \geq 0 : |Y(s)| \geq n\}$, $\tau_Z = \inf\{s \geq 0 : |Z(s))| \geq n\}$ and $\tau = \tau_Y \wedge \tau_Z$. Let F denote the $\mathcal{F}_0$-measurable set $\{Y(0) = Z(0)\}$. Then $Y(\cdot \wedge \tau)$ and $Z(\cdot \wedge \tau)$ are indistinguishable on F, i.e.,

$$P(\{Y(t \wedge \tau) \neq Z(t \wedge \tau) \text{ for some } t \geq 0\} \cap F) = 0.$$

Proof. By subtracting (10.30) from (10.29) and stopping at the time τ, we obtain for all $t \geq 0$,

$$
\begin{aligned}
Y(t \wedge \tau) &- Z(t \wedge \tau) \\
&= Y(0) - Z(0) + \int_0^{t \wedge \tau} (\sigma_Y(Y(s)) - \sigma_Z(Z(s)))dB(s) \\
&\quad + \int_0^{t \wedge \tau} (b_Y(Y(s)) - b_Z(Z(s)))ds.
\end{aligned}
$$

(10.31)

The integrals in (10.31) are unchanged if the integrands are multiplied by $1_{(0, t \wedge \tau]}$ (cf. Theorem 2.7(iii) and (2.11)). Moreover, since Y and Z are continuous, $Y(s)$ and $Z(s)$ are in $\{x \in \mathbb{R}^d : |x| \leq n\}$ for $0 < s \leq \tau$. Hence, on F we have

$$Y(t \wedge \tau) - Z(t \wedge \tau) = \int_0^t 1_{(0,t\wedge\tau]}(s)(\sigma(Y(s)) - \sigma(Z(s)))dB(s)$$
$$+ \int_0^t 1_{(0,t\wedge\tau]}(s)(b(Y(s)) - b(Z(s)))ds.$$

Then it follows in a similar manner to the proof of Lemma 10.1, with $1_{(0,t\wedge\tau]}$ and 1_F inserted in all integrals there, and (10.25)–(10.28) used in place of (10.6)–(10.7), that

$$E\left[\sup_{0 \le s \le t} |Y(s \wedge \tau) - Z(s \wedge \tau)|^2; F\right]$$
$$\le 2K_n^2(4r + t)E\left[\int_0^t |Y(s \wedge \tau) - Z(s \wedge \tau)|^2 ds; F\right].$$

Here $E[\,\cdot\,; F]$ denotes expectation taken over the set F. Then, by Gronwall's inequality (Lemma 10.2),

$$E[|Y(t \wedge \tau) - Z(t \wedge \tau)|^2; F] = 0.$$

Hence $Y(t \wedge \tau) = Z(t \wedge \tau)$ P-a.s. on F for each $t \in \mathbb{R}_+$. But since Y and Z have continuous paths, it follows that $Y(\cdot \wedge \tau)$ and $Z(\cdot \wedge \tau)$ are indistinguishable on F. ∎

Proof of Theorem 10.6. The uniqueness in Theorem 10.6 follows from Lemma 10.7. For if Y and Z are two continuous adapted solutions of (10.4), then for each $n \in \mathbb{N}$ and $\tau_n \equiv \inf\{t \ge 0 : |Y(t)| \ge n \text{ or } |Z(t)| \ge n\}$, $Y(\cdot \wedge \tau_n)$ and $Z(\cdot \wedge \tau_n)$ are indistinguishable. But then

$$\tau_n = \inf\{t \ge 0 : |Y(t)| \ge n\} = \inf\{t \ge 0 : |Z(t)| \ge n\}, \quad P\text{-a.s.}$$

By continuity of paths, $\tau_n \uparrow \infty$ P-a.s. as $n \to \infty$, and then it follows that Y and Z are indistinguishable.

For the proof of existence, we first show that it suffices to consider the case where $X(0)$ is bounded. For if $X(0)$ is not bounded, and we assume Theorem 10.6 holds for the case of bounded $X(0)$, then for each positive integer n we can define a process $\bar{X}^n$ to be the solution of (10.4) with

$$\bar{X}^n(0) = X(0)1_{\{|X(0)| < n\}} + n1_{\{|X(0)| \ge n\}}.$$

By Lemma 10.7 with $\sigma_Y = \sigma_Z = \sigma$, $b_Y = b_Z = b$, we have for all $m \geq n$ that $\bar{X}^m$ and $\bar{X}^n$ are indistinguishable on $\{|X(0)| < n\}$. But it then follows in a similar manner to that in Section 2.6 that there is a continuous adapted process $\bar{X}$ such that P-a.s., $\bar{X}(t) = \lim_{n \to \infty} \bar{X}^n(t)$ for all $t \geq 0$ and $\bar{X}$ satisfies (10.4). Indeed, $\bar{X} = \bar{X}^n$, P-a.s. on $\{|X(0)| < n\}$. Moreover, since the law of $\bar{X}^n$ is uniquely determined by the laws of $X(0)$ and B, the same is true for $\bar{X}$. Thus it remains to consider the case of a bounded $X(0)$. In the following we will assume the weaker condition that $X(0) \in L^2$, because the details of the proof will be used later in this chapter.

To prove existence for $X(0) \in L^2$, we first construct solutions with bounded coefficients that approximate σ and b. Thus for each $n \in I\!N$, we let $\sigma_n : I\!R^d \to I\!R^d \otimes I\!R^r$ be a bounded, uniformly Lipschitz function such that $\sigma_n = \sigma$ on $\{x \in I\!R^d : |x| \leq n\}$. Such a σ_n can be obtained by multiplying σ by a real-valued function that takes the value one on $\{x : |x| \leq n\}$, is zero on $\{x : |x| \geq 2n\}$, and is uniformly Lipschitz in between. Similarly, let $b_n : I\!R^d \to I\!R^d$ be a bounded, uniformly Lipschitz function such that $b_n = b$ on $\{x \in I\!R^d : |x| \leq n\}$.

By Theorem 10.5, for each $n \in I\!N$ there is a unique solution X^n of (10.4) when the coefficients σ, b there are replaced by σ_n, b_n. Then, by Lemma 10.7 with $Y = X^m$, $Z = X^n$ and $m \geq n$, we have $X^m(\cdot \wedge \tau_n)$ and $X^n(\cdot \wedge \tau_n)$ are indistinguishable, where P-a.s.

$$\tau_n \equiv \inf\{t \geq 0 : |X^n(t)| \geq n\} = \inf\{t \geq 0 : |X^m(t)| \geq n\}.$$

If we can prove that $\tau_n \uparrow \infty$ P-a.s. as $n \to \infty$, then it will follow that there is a continuous adapted process X such that P-a.s., $X(t) = \lim_{n \to \infty} X^n(t)$ for all $t \geq 0$, and in particular, P-a.s., $X(t) = X^n(t)$ for all $t \in [0, \tau_n)$. It will then follow that X satisfies (10.4) (see Exercise 3).

To prove $\tau_n \uparrow \infty$ P-a.s. as $n \to \infty$, we apply Itô's formula to $\phi(x) = |x|^2$ and $X^n(\cdot \wedge \tau_n)$, and we use the facts that $\sigma_n(X_s^n) = \sigma(X_s^n)$ and $b_n(X_s^n) = b(X_s^n)$ for $s \in (0, t \wedge \tau_n]$, to obtain

$$|X^n(t \wedge \tau_n)|^2 - |X^n(0)|^2$$

(10.32)
$$= 2 \sum_{i=1}^{d} \sum_{j=1}^{r} \int_0^{t \wedge \tau_n} X_i^n(s) \sigma_{ij}(X^n(s)) dB_j(s)$$

$$+ 2 \sum_{i=1}^{d} \int_0^{t \wedge \tau_n} X_i^n(s) b_i(X^n(s)) ds + \sum_{i=1}^{d} \sum_{j=1}^{r} \int_0^{t \wedge \tau_n} \sigma_{ij}^2(X^n(s)) ds.$$

Since $E(|X^n(0)|^2) = E(|X(0)|^2) < \infty$ and the integrands in all of the integrals are bounded on $(0, t \wedge \tau_n]$, we can take expectations in the above. Noting that the stochastic integrals are martingales (and so have zero expectation), we obtain

$$E(|X^n(t \wedge \tau_n)|^2) - E(|X(0)|^2)$$

(10.33)
$$= 2E\left(\int_0^{t \wedge \tau_n} X^n(s) \cdot b(X^n(s)) ds \right) + E\left(\int_0^{t \wedge \tau_n} \|\sigma(X^n(s))\|^2 ds \right).$$

By the growth conditions on σ, b, we deduce from (10.33) that

$$E(|X^n(t \wedge \tau_n)|^2) \le E(|X(0)|^2) + 3KE\left(\int_0^{t \wedge \tau_n} (1 + |X^n(s)|^2) ds \right).$$

Then

$$E(1 + |X^n(t \wedge \tau_n)|^2) \le 1 + E(|X(0)|^2) + 3K \int_0^t E(1 + |X^n(s \wedge \tau_n)|^2) ds.$$

It follows from Gronwall's inequality (Lemma 10.2) that for $C_1 = 1 + E(|X(0)|^2) < \infty$,

(10.34)
$$E(|X^n(t \wedge \tau_n)|^2) \le C_1 e^{3Kt} \equiv C_2(t).$$

Hence, by the definition of τ_n,

(10.35)
$$P(0 < \tau_n \le t) n^2 \le C_2(t),$$

and so

$$P(\tau_n \le t) = P(\tau_n = 0) + P(0 < \tau_n \le t)$$
$$\le P(|X(0)| \ge n) + C_2(t)n^{-2}$$
$$\to 0 \text{ as } n \to \infty.$$

But $\{\tau_n\}$ is increasing and t is arbitrary, so we conclude that P-a.s., $\tau_n \uparrow \infty$ as $n \to \infty$, as required.

By Theorem 10.5, the law of X^n and hence of $X^n(\cdot \wedge \tau_n)$ is uniquely determined by σ_n, b_n and the laws of $X(0)$ and B. But by Lemma 10.7, the dependence of $X^n(\cdot \wedge \tau_n)$ on σ_n and b_n is only through their values in $\{x \in \mathbb{R}^d : |x| \le n\}$, where they agree with σ and b. Hence the law of $X^n(\cdot \wedge \tau_n)$ is uniquely determined by σ, b and the laws of $X(0)$ and B. Since $X(\cdot) = \lim_{n \to \infty} X^n(\cdot \wedge \tau_n)$ P-a.s., it follows that the law of X is uniquely determined by these same things. $\blacksquare$

Corollary 10.8. *Suppose all of the hypotheses of Theorem 10.6 hold and in addition that $X(0) \in L^2$. Then $X(t) \in L^2$ for all $t \ge 0$.*

Proof. This follows immediately on letting $n \to \infty$ in (10.34) and invoking Fatou's lemma. $\blacksquare$

Extensions.

1. To simplify the exposition, we have only considered time-homogeneous coefficients σ and b here. The results of Theorem 10.6 can be readily extended to the situation where σ and b also depend on time, provided these coefficients are continuous and the Lipschitz and growth conditions (10.25)–(10.28) hold uniformly on compact time intervals. One can even allow $\sigma(t, \cdot)$ and $b(t, \cdot)$ to depend on the history of the path up to time t. Conditions (10.25)–(10.28) have then to be modified to reflect Lipschitz and growth conditions on bounded subsets of path space that are uniform for t's in compact time intervals (see Revuz and Yor [64, §V.3] or Rogers and Williams [68, V.11–V.12] for details).

2. In the special case when $d = 1$, the Lipschitz condition on σ can be weakened to a Hölder condition. In particular, if σ is locally Hölder continuous with exponent $\frac{1}{2}$ and conditions (10.26)–(10.28) hold, then Theorem 10.6

still holds (see Ikeda and Watanabe [45, p. 168] or Karatzas and Shreve [49, p. 291]).

10.3 Strong Markov Property of the Solution

As in the last section, suppose σ and b satisfy (10.25)–(10.28) and B is an r-dimensional Brownian motion martingale starting from the origin, defined on some filtered probability space $(\Omega, \mathcal{F}, \{\mathcal{F}_t\}, P)$.

For each $x \in I\!\!R^d$, let X^x denote the unique solution of (10.4) satisfying $X(0) = x$. Note that all these solutions for varying x can be realized on a single filtered probability space on which B is a Brownian motion martingale starting from the origin, since the deterministic initial random variable x is $\mathcal{F}_0$-measurable for all $x \in I\!\!R^d$. By Theorem 10.6, the law of X^x is uniquely determined by x and the law of B. Thus any two realizations on different probability spaces will have the same law. A canonical realization of this law can be obtained by considering the probability measure P^x induced on the space of continuous paths in $I\!\!R^d$ by X^x and P. Specifically, let

$$C_{I\!\!R^d} = \{w : [0, \infty) \to I\!\!R^d, \ w \text{ is continuous}\}$$

with the natural σ-field $\mathcal{M} = \sigma\{w(s) : 0 \leq s < \infty\}$ and filtration $\{\mathcal{M}_t, t \geq 0\}$ for $\mathcal{M}_t = \sigma\{w(s) : 0 \leq s \leq t\}$. Note that we have not completed the σ-field $\mathcal{M}$ nor augmented $\{\mathcal{M}_t, t \geq 0\}$ at this point (cf. the last paragraph of this section). For each $x \in I\!\!R^d$, let P^x denote the probability measure defined on $(C_{I\!\!R^d}, \mathcal{M})$ by

$$P^x(A) = P(X^x(\cdot) \in A) \text{ for all } A \in \mathcal{M}.$$

The expectation with respect to P^x will be denoted by E^x. We will show that the canonical process $w \to w(\cdot)$, together with the probability measures $\{P^x, x \in I\!\!R^d\}$, defines a continuous strong Markov process. The proof of this is a rather delicate matter frequently glossed over in other texts. We state and prove two lemmas (10.9 and 10.10) to make the requisite steps perfectly clear. These lemmas figure in the proof of the Markov property in showing that the right member of (10.43) is an $\mathcal{M}_s$-measurable random variable and that it has the same distribution as the left member. In addition, Lemma 10.9 plays a crucial role in the extension of the Markov property to the strong Markov property (see Theorem 10.12).

Let $C_b(\mathbb{R}^d)$ denote the set of bounded continuous real-valued functions on $\mathbb{R}^d$. For $f \in C_b(\mathbb{R}^d)$, let $||f|| = \sup\limits_{x \in \mathbb{R}^d} |f(x)|$.

Lemma 10.9. *For each $f \in C_b(\mathbb{R}^d)$ and $t \geq 0$, $E^x[f(w(t))]$ is a continuous function of $x \in \mathbb{R}^d$.*

Remark. This is a variation of the Feller property as defined in Chung [12].

Proof. It is sufficient to prove that $x \to E^x[f(w(t))]$ is continuous in each ball centered at the origin. Let $f \in C_b(\mathbb{R}^d)$, $t \geq 0$, $R > 0$ and $U_R = \{x \in \mathbb{R}^d : |x| \leq R\}$. Let $x, y \in U_R$ and define $\tau_n^z = \inf\{t \geq 0 : |X_t^z| \geq n\}$ for $z = x, y$, and $n \in \mathbb{N}$. Then we have

$$|E^x[f(w(t))] - E^y[f(w(t))]|$$
$$= |E[f(X_t^x) - f(X_t^y)]|$$
$$\leq |E[f(X_t^x) - f(X_t^y); \tau_n^x > t, \tau_n^y > t]| + 2||f||(P(\tau_n^x \leq t) + P(\tau_n^y \leq t)).$$

Let $X^{x,n}$ and $X^{y,n}$ denote the approximations (with bounded coefficients σ_n, b_n) to X^x and X^y used in the proof of Theorem 10.6. Since these approximations agree with X^x and X^y until the times τ_n^x and τ_n^y respectively, it follows that the last line above equals

(10.36)
$$|E[f(X_t^{x,n}) - f(X_t^{y,n}); \tau_n^x > t, \tau_n^y > t]|$$
$$+ 2||f||(P(\tau_n^x \leq t) + P(\tau_n^y \leq t)).$$

Now for $n > R$, $\tau_n^z > 0$ for all $z \in U_R$ and then by (10.35),

$$P(\tau_n^z \leq t) \leq C_2(t)n^{-2},$$

where $C_2(t) = (1+|z|^2)e^{3Kt}$ does not depend on n and is uniformly bounded for $z \in U_R$. Thus, given $\varepsilon > 0$, for all $n > R$ sufficiently large we have

$P(\tau_n^z \le t) < \varepsilon/8(\|f\|+1)$ for all $z \in U_R$. Fixing such an n, (10.36) is bounded by

$$(10.37) \qquad |E[f(X_t^{x,n}) - f(X_t^{y,n}); \tau_n^x > t, \tau_n^y > t]| + \frac{\varepsilon}{2}.$$

Now f is uniformly continuous on $U_n = \{z \in \mathbb{R}^d : |z| \le n\}$ and so there is $\eta > 0$ such that

$$(10.38) \qquad |f(z) - f(z')| < \frac{\varepsilon}{4} \text{ whenever } |z - z'| < \eta \text{ and } z, z' \in U_n.$$

By applying Lemma 10.1 with $\tilde{Y} = Y = X^{x,n}$ and $\tilde{Z} = Z = X^{y,n}$, and then Lemma 10.2, we obtain

$$E[|X_t^{x,n} - X_t^{y,n}|^2] \le 3|x - y|^2 e^{C_t t},$$

where C_t depends on the Lipschitz constants of σ_n, b_n, and on t, but not on x or y. Then, by Chebyshev's inequality, there is $\delta > 0$ such that

$$(10.39) \qquad P(|X_t^{x,n} - X_t^{y,n}| \ge \eta) < \frac{\varepsilon}{8(\|f\|+1)}$$

whenever $|x - y| < \delta$. Combining (10.38)–(10.39), and noting that $X_t^{z,n} \in U_n$ when $\tau_n^z > t$, for $z = x, y$, we see that (10.37) is bounded by ε whenever $|x - y| < \delta$. Since $\varepsilon > 0$ was arbitrary, the continuity of $x \to E^x[f(w(t))]$ on U_R follows. ∎

Lemma 10.10. *Let X be a solution of (10.4). Then for any $f \in C_b(\mathbb{R}^d)$ and $t \ge 0$ we have*

$$(10.40) \qquad E[f(X_t)|X_0] = E^{X_0}[f(w_t)].$$

Proof. Observe that by Lemma 10.9, the right member above is measurable with respect to the σ-field generated by X_0.

We first show that it suffices to consider a bounded $X(0)$. For this, suppose $X(0)$ is unbounded and that Lemma 10.10 holds for bounded $X(0)$. For each positive integer n, define

$$\tilde{X}^n(0) = X(0)1_{\{|X(0)|<n\}} + n1_{\{|X(0)|\ge n\}},$$

and let $\bar{X}^n$ denote the solution of (10.4) with initial value $\bar{X}^n(0)$. By the proof of Theorem 10.6, on $\{|X(0)| < n\}$ we have P-a.s.

$$(10.41) \qquad X(t) = \bar{X}^n(t) \quad \text{for all } t \geq 0.$$

Moreover, by applying (10.40) to $\bar{X}^n$ we obtain

$$(10.42) \qquad E[f(\bar{X}^n(t)) \mid \bar{X}_0^n] = E^{\bar{X}_0^n}[f(w_t)].$$

Hence, for any Borel set A in $I\!\!R^d$, we have

$$
\begin{aligned}
\int_{\{X_0 \in A\}} f(X_t)\, dP &= \lim_{n \to \infty} \int_{\{X_0 \in A\}} f(X_t) 1_{\{|X_0| < n\}}\, dP \\
&= \lim_{n \to \infty} \int f(\bar{X}_t^n) 1_{\{\bar{X}_0^n \in A,\, |\bar{X}_0^n| < n\}}\, dP \\
&= \lim_{n \to \infty} \int E^{\bar{X}_0^n}[f(w_t)] 1_{\{\bar{X}_0^n \in A,\, |\bar{X}_0^n| < n\}}\, dP \\
&= \lim_{n \to \infty} \int E^{X_0}[f(w_t)] 1_{\{X_0 \in A,\, |X_0| < n\}}\, dP \\
&= \int_{\{X_0 \in A\}} E^{X_0}[f(w_t)]\, dP,
\end{aligned}
$$

where we have used (10.41) for the second and fourth equalities and (10.42) for the third equality. It follows that the lemma holds for unbounded X_0, given that it holds for bounded X_0.

We now consider the case where X_0 is bounded. For each positive integer k, let Y_0^k be an $\mathcal{F}_0$-measurable random variable taking countably many values in $I\!\!R^d$ such that $\|X_0 - Y_0^k\|_\infty < 1/k$, where $\|\cdot\|_\infty$ is the L^∞-norm on $(\Omega, \mathcal{F}, P)$. Let Y^k be the continuous adapted solution of (10.4) with Y_0^k in place of $X(0)$, the existence and uniqueness of which is guaranteed by Theorem 10.6. As in the proof of that theorem, let σ_n, b_n be bounded, uniformly Lipschitz functions that agree with σ, b on $U_n = \{x \in I\!\!R^d : |x| \leq n\}$. Let $X^n, Y^{k,n}$ be the solutions of (10.4) that arise with these coefficients in place of σ, b and with initial conditions $X^n(0) = X_0, Y^{k,n}(0) = Y_0^k$, respectively. Define $\tau_n = \inf\{t \geq 0 : |X^n(t)| \geq n\}$ and $\tau_n^k = \inf\{t \geq 0 : |Y^{k,n}(t)| \geq n\}$.

Then, as in the proof of Lemma 10.9, we have for each $f \in C_b(\mathbb{R}^d)$ and $t \geq 0$,

$$E[|f(X_t) - f(Y_t^k)|] \leq E[|f(X_t^n) - f(Y_t^{k,n})|; \tau_n > t, \tau_n^k > t]$$
$$+ 2\|f\|(P(\tau_n \leq t) + P(\tau_n^k \leq t)).$$

Here, by (10.35) and the fact that $\|X_0 - Y_0^k\|_\infty < k^{-1} \leq 1$, we have

$$P(\tau_n^k \leq t) = P(\tau_n^k = 0) + P(0 < \tau_n^k \leq t)$$
$$= P(|Y_0^k| \geq n) + (1 + E(|Y_0^k|^2))e^{3Kt} n^{-2}$$
$$\leq P(|X_0| \geq n - 1) + (1 + E((|X_0| + 1)^2))e^{3Kt} n^{-2}.$$

The last line tends to zero uniformly in k as $n \to \infty$. Similarly, by (10.35), $P(\tau_n \leq t) \to 0$ as $n \to \infty$. Thus, given $\varepsilon > 0$, by choosing n sufficiently large we can ensure that

$$2\|f\|(P(\tau_n \leq t) + P(\tau_n^k \leq t)) < \frac{\varepsilon}{2}$$

for all k. Fixing such an n, by Lemmas 10.1 and 10.2 we have

$$E[|X_t^n - Y_t^{k,n}|^2] \leq 3E[|X_0 - Y_0^k|^2]e^{C_t t} \leq 3k^{-2}e^{C_t t},$$

where C_t depends on the Lipschitz constants of σ_n, b_n and on t. Then, in the same manner as for Lemma 10.9, we can use the above, together with the uniform continuity of f on U_n, to deduce that there is an integer $k(\varepsilon)$ such that

$$E[|f(X_t^n) - f(Y_t^{k,n})|; \tau_n > t, \tau_n^k > t] < \frac{\varepsilon}{2} \quad \text{for all } k \geq k(\varepsilon).$$

Hence

$$E[|f(X_t) - f(Y_t^k)|] < \varepsilon \text{ for } k \geq k(\varepsilon).$$

Since $\varepsilon > 0$ was arbitrary, we have $f(Y_t^k) \to f(X_t)$ in $L^1(\Omega, \mathcal{F}, P)$ as $k \to \infty$. Thus, for $g \in C_b(\mathbb{R}^d)$,

$$E[g(X_0)f(X_t)] = \lim_{k \to \infty} E[g(Y_0^k)f(Y_t^k)].$$

Enumerating the countably-many values taken by Y_0^k as $\{y_j^k\}$, the right member above is equal to

$$\lim_{k\to\infty} \sum_j E[g(y_j^k)1_{\{Y_0^k=y_j^k\}}f(Y_t^k)]$$

$$= \lim_{k\to\infty} \sum_j E[g(y_j^k)1_{\{Y_0^k=y_j^k\}}E[f(Y_t^k)|Y_0^k = y_j^k]].$$

But conditioned on the non-trivial event $\{Y_0^k = y_j^k\}$, $Y^k(\cdot)$ is a solution of (10.4) satisfying $Y^k(0) = y_j^k$. By the uniqueness in law of such a solution, this conditioned process has the same law as $w(\cdot)$ under $P^{y_j^k}$. Thus the right member above equals

$$\lim_{k\to\infty} \sum_j E\left[g(y_j^k)1_{\{Y_0^k=y_j^k\}}E^{y_j^k}[f(w(t))]\right] = \lim_{k\to\infty} E[g(Y_0^k)E^{Y_0^k}[f(w(t))]].$$

By the continuity of g and $x \to E^x[f(w(t))]$, the limit above equals $E[g(X_0)E^{X_0}[f(w(t))]]$, by bounded convergence. Thus

$$E[g(X_0)f(X_t)] = E\left[g(X_0)E^{X_0}[f(w_t)]\right] \quad \text{for all } g \in C_b(\mathbb{R}^d).$$

Since $E^{X_0}[f(w_t)]$ is measurable with respect to the σ-field generated by X_0, and the latter is generated by $\{g(X_0) : g \in C_b(\mathbb{R}^d)\}$, the desired result follows. ∎

Now we verify the Markov property of the process $w(\cdot)$ on $(C_{\mathbb{R}^d}, \mathcal{M}, \{\mathcal{M}_t\})$ with the associated family of measures $\{P^x, x \in \mathbb{R}^d\}$. In the notation of Blumenthal and Getoor [5] or Sharpe [71], this canonical representation of the process would be denoted by

$$(C_{\mathbb{R}^d}, \mathcal{M}, \mathcal{M}_t, w(t), \theta_t, P_x),$$

where θ_t denotes the usual shift operator on $C_{\mathbb{R}^d}$ given by $(\theta_t w)(\cdot) = w(\cdot + t)$.

Theorem 10.11. Let $x \in \mathbb{R}^d$, $f \in C_b(\mathbb{R}^d)$, and $s, t \in \mathbb{R}_+$. Then we have P^x-a.s.,

$$(10.43) \qquad E^x[f(w(t+s))|\mathcal{M}_s] = E^{w(s)}[f(w(t))].$$

Remark. This is not the first time that we have abused notation by using the same variable w in different contexts within the same formula. One could avoid this ambiguity by replacing the last w by $\tilde{w}$ to denote a generic element of $C_{I\!\!R^d}$.

Proof. Note that the right member of (10.43) is well defined as an $\mathcal{M}_s$-measurable random variable by Lemma 10.9 and composition of mappings. By shifting time by s in the integrals in the stochastic equation (10.4) for X^x, we obtain P-a.s.,

$$X^x_{t+s} = X^x_s + \int_0^t \sigma(X^x_{u+s})dB_{u+s} + \int_0^t b(X^x_{u+s})du.$$

Let $\mathcal{G}$ denote the completion of the σ-algebra generated by X^x_s and $B_{\cdot+s} - B_s$, and for each $t \geq 0$ let $\mathcal{G}_t$ denote the σ-field generated by X^x_s and $\{B_{u+s} - B_s : 0 \leq u \leq t\}$, augmented by the P-null sets in $\mathcal{G}$. Then, by the above, $\bar{X}_{\cdot} = X^x_{\cdot+s}$ is a solution on the filtered probability space $(\Omega, \mathcal{G}, \{\mathcal{G}_t\}, P)$ of the stochastic differential equation

(10.44) $$d\bar{X}(t) = \sigma(\bar{X}(t))d\bar{B}(t) + b(\bar{X}(t))dt,$$

with initial condition $\bar{X}(0) = X^x_s$ and $\bar{B}(\cdot) = B(\cdot + s) - B(s)$. (Note the stochastic integral is the same if $\bar{B}_{\cdot}$ is used in place of $B_{\cdot+s}$.) By Theorem 10.6, this solution is unique. Moreover, by the construction in the proof of that theorem which is valid for $X(0) \in L^2$, if we let $\bar{X}^n$ denote the solution of (10.44) with σ_n, b_n in place of σ, b, respectively, then P-a.s.

$$\bar{X}(t) = \lim_{n\to\infty} \bar{X}^n(t) \quad \text{for all } t \geq 0.$$

By the construction in Theorem 10.4, for each n we have P-a.s.

$$\bar{X}^n(t) = \lim_{m\to\infty} \bar{X}^{n,m}(t) \quad \text{for all } t \geq 0,$$

where $\bar{X}^{n,0} = \bar{X}(0)$ and

$$\bar{X}^{n,m} = \bar{X}(0) + \int_0^t \sigma_n\left(\bar{X}^{n,m-1}(s)\right)d\bar{B}(s) + \int_0^t b_n\left(\bar{X}^{n,m-1}(s)\right)ds$$

for all $m \geq 1$. By representing the integrals in the above as limits of sums, it can be shown by induction on m, using the independence of $\bar{B}$ from $\mathcal{F}_s$, that the law of $\bar{X}^{n,m}$ conditioned on $\mathcal{F}_s$ is the same as the law of $\bar{X}^{n,m}$ conditioned on $\bar{X}(0)$. It follows from this and bounded convergence that

$$
\begin{aligned}
E\left[f\left(X_{t+s}^x\right)\mid\mathcal{F}_s\right] &= E\left[f\left(\bar{X}_t\right)\mid\mathcal{F}_s\right] \\
&= \lim_{n\to\infty}\lim_{m\to\infty} E\left[f\left(\bar{X}_t^{n,m}\right)\mid\mathcal{F}_s\right] \\
&= \lim_{n\to\infty}\lim_{m\to\infty} E\left[f\left(\bar{X}_t^{n,m}\right)\mid\bar{X}(0)\right] \\
&= E\left[f\left(\bar{X}_t\right)\mid\bar{X}(0)\right].
\end{aligned}
$$

But then by Lemma 10.10 applied to $\bar{X}$ with initial value $\bar{X}(0) = X_s^x$, we see that the last member above equals $E^{X_s^x}[f(w_t)]$. Thus

$$
E\left[f\left(X_{t+s}^x\right)\mid\mathcal{F}_s\right] = E^{X_s^x}[f(w_t)].
$$

We claim that (10.43) now follows. To see this, let $k\in\mathbb{N}$, $g_1,\ldots,g_k\in C_b(\mathbb{R}^d)$, and $0\le s_1<\ldots<s_k\le s$. Then by the definition of P^x,

$$
\begin{aligned}
E^x&[f(w(t+s))g_1(w(s_1))\ldots g_k(w(s_k))] \\
&= E[f(X_{t+s}^x)g_1(X_{s_1}^x)\ldots g_k(X_{s_k}^x)] \\
&= E[E[f(X_{t+s}^x)\mid\mathcal{F}_s]g_1(X_{s_1}^x)\ldots g_k(X_{s_k}^x)] \\
&= E[E^{X_s^x}[f(w(t))]g_1(X_{s_1}^x)\ldots g_k(X_{s_k}^x)] \\
&= E^x[E^{w(s)}[f(w(t))]g_1(w(s_1))\ldots g_k(w(s_k))].
\end{aligned}
$$

Since functions of the form $w\to g(w(s_1))\ldots g(w(s_k))$ generate $\mathcal{M}_s$, (10.43) follows. ∎

Theorem 10.12. *For each $x\in\mathbb{R}^d$, $f\in C_b(\mathbb{R}^d)$, and each optional time τ (relative to $\{\mathcal{M}_t\}$), we have for $t\ge 0$:*

(10.45) $$E^x[f(w_{\tau+t})\mid\mathcal{M}_{\tau+}] = E^{w_\tau}[f(w_t)] \quad\text{on } \{\tau<\infty\}.$$

Proof. This follows from Lemma 10.9 and Theorem 10.12 in the same manner as Theorem 1 of §2.3 of Chung [12]. Note that only the property in Lemma 10.9 is needed for that proof. ∎

Property (10.45) is one of the equivalent forms of the strong Markov property for the canonical process $w(\cdot)$ under $\{P^x\}_{x\in\mathbb{R}^d}$. In the usual manner (see §2.3 of Chung [12]), by augmentation, the family $\{\mathcal{M}_t\}_{t \geq 0}$ can be made right continuous so that $\tau+$ in (10.45) can then be replaced by τ. One can obtain the law of a solution of (10.4) with any initial distribution by randomizing over the P^x's.

10.4 Strong and Weak Solutions

In this section, we consider the two basic kinds of solution of the SDE (10.1) and discuss related notions of uniqueness.

Although there is common agreement as to what one means by a *weak* solution of (10.1), there are at least two interpretations of what is meant by a *strong* solution. Loosely speaking, a strong solution of (10.1) is a solution that is adapted to the filtration generated by B and $X(0)$. The disagreement comes in the way this is made precise. In Durrett [27] and Karatzas and Shreve [49], for example, a strong solution is defined relative to a given probability space, Brownian motion B and independent initial random variable $X(0)$. On the other hand, in Ikeda and Watanabe [45, p. 149] and Rogers and Williams [68, §V.10], a strong solution is a suitably measurable *function* $F : \mathbb{R}^d \times C_{\mathbb{R}^r} \to C_{\mathbb{R}^d}$, where

$$C_{\mathbb{R}^n} = \{w : [0, \infty) \to \mathbb{R}^n, \, w \text{ is continuous}\}, \quad n \geq 1,$$

such that for any Brownian motion B and independent initial random variable $X(0)$, $X = F(X(0), B)$ is a strong solution of (10.1) in the sense of [27] and [49]. The measurability assumptions imposed on F by [68] and [45] include joint measurability in $(x, w) \in \mathbb{R}^d \times C_{\mathbb{R}^d}$. This is the feature that distinguishes their notions of strong solution from that of [27] and [49]. This definition of a strong solution by means of a function F is most useful for studying stochastic flows, where one is interested in the simultaneous movement of all points in the state space under the action of the stochastic differential equation (10.1). We shall not pursue the subject of stochastic flows here, but refer the interested reader to Rogers and Williams [68, §V.13 ff.]. If the function F is continuous in x, it can be used in verifying the strong Markov property (cf. Lemma 10.9), but the latter can also be proved by alternative means, e.g., via uniqueness of solutions to martingale problems introduced below in the discussion of weak solutions.

Here we shall adopt the approach that the terms weak and strong solution refer to the stochastic differential with some specific data, whereas broad notions of existence and uniqueness will relate to the formal SDE (10.1). We propose definitions that convey the appropriate intuitive meaning and are at least consistent within the confines of this book. This is one area where one cannot be consistent with all pre-existing terminology.

Definition. Let $(\Omega, \mathcal{F}, P)$ be a complete probability space. Suppose that there are defined on this space a d-dimensional Brownian motion B that starts from the origin, and an $I\!\!R^d$-valued random variable ξ such that ξ is independent of B. Let $\{\mathcal{F}_t, t \geq 0\}$ be the standard filtration generated by ξ and B, i.e.,

$$\mathcal{F}_t = \sigma\{\xi; B(s) : 0 \leq s \leq t\}^{\sim},$$

where $\sim$ denotes augmentation by all of the P-null sets in $\mathcal{F}$. Note that B is a martingale relative to $\{\mathcal{F}_t\}$. A *strong solution* of (10.1) with the data $(\Omega, \mathcal{F}, P, B, \xi)$ is a continuous d-dimensional process X defined on $(\Omega, \mathcal{F}, P)$ such that the following four conditions hold.

(i) X is adapted to the filtration $\{\mathcal{F}_t\}$.

(ii) $X(0) = \xi$ P-a.s.

(iii) $\int_0^t \{|b(X(s))| + \|\sigma(X(s))\|^2\} ds < \infty$ P-a.s. for each $t \geq 0$.

(iv)   Equation (10.4) holds P-a.s. for all $t \geq 0$.

Remarks.

1. Since X is continuous and adapted, and σ and b are Borel measurable, $\sigma(X)$ and $b(X)$ are $\mathcal{B} \times \mathcal{F}$-measurable. Then by Theorem 3.8 and Lemma 3.11, together with condition (iii) above, $\int_0^t \sigma(X_s) dB_s$ defines a continuous local martingale and $\int_0^t b(X_s) ds$ defines a continuous adapted process that is locally of bounded variation.

2. In property (iv) above (and subsequently in the definition of a weak solution), the order of the qualifiers "P-a.s." and "for all $t \geq 0$" can be interchanged, since both sides of (10.4) are continuous in t.

Definition. We say *strong existence* holds for the SDE (10.1) if for each quintuple $(\Omega, \mathcal{F}, P, B, \xi)$ satisfying the conditions of the above definition, there is a strong solution of (10.1) with this data. We say *strong uniqueness* holds for the SDE (10.1) if, for each possible quintuple of data, there is at most one strong solution of (10.1).

Remark. It follows from Theorem 10.6 that when σ and b satisfy (10.25)–(10.28), strong existence and uniqueness hold for (10.1).

Example. *(Ornstein-Uhlenbeck process.)* Recall the Langevin stochastic differential equation discussed in Chapter 5 in connection with the Ornstein-Uhlenbeck process,

$$(10.46) \qquad dX_t = -\alpha X_t\, dt + dB_t.$$

Since the coefficients $\sigma(x) \equiv 1$ and $b(x) = -\alpha x$ satisfy (10.25)–(10.28), we have strong existence and uniqueness for this equation. Indeed, one can verify this directly, as follows.

Given a quintuple of data $(\Omega, \mathcal{F}, P, B, \xi)$, as in the above definition of a strong solution, define

$$(10.47) \qquad X_t = e^{-\alpha t}\xi + e^{-\alpha t}\int_0^t e^{\alpha s}\, dB_s, \quad \text{for all } t \geq 0.$$

Then, by the analysis in Chapter 5, X is a solution of (10.46), and by its definition, it is adapted to the standard filtration generated by ξ and B. Hence X is a strong solution of (10.46) with the given data. Since the data was arbitrary, it follows that strong existence holds for (10.46).

On the other hand, for any strong solution X of (10.46) with associated data $(\Omega, \mathcal{F}, P, B, \xi)$, we have by Itô's formula:

$$e^{\alpha t} X_t - X_0 = \int_0^t e^{\alpha s}(-\alpha X_s)\, ds + \int_0^t e^{\alpha s}\, dB_s + \alpha \int_0^t e^{\alpha s} X_s\, ds$$

$$= \int_0^t e^{\alpha s}\, dB_s.$$

Setting $X_0 = \xi$ and multiplying by $e^{-\alpha t}$, we see that X is of the form (10.47). Thus, strong uniqueness holds.

Definition. Let μ be a Borel probability measure on $\mathbb{R}^d$. A *weak solution* of (10.1) with initial law μ is a sextuple $(\Omega, \mathcal{F}, \{\mathcal{F}_t\}, P, B, X)$ such that

$(\Omega, \mathcal{F}, \{\mathcal{F}_t\}, P)$ is a filtered probability space, and B and X are processes defined on this space satisfying the following four conditions.

(i) B is a continuous r-dimensional Brownian motion martingale and X is a continuous adapted d-dimensional process.

(ii) $X(0)$ has distribution μ.

(iii) $\int_0^t \left\{ |b(X(s))| + \|\sigma(X(s))\|^2 \right\} ds < \infty$ P-a.s. for all $t \geq 0$.

(iv)   Equation (10.4) holds P-a.s. for all $t \geq 0$.

Remarks.

1. Note that in specifying a weak solution one must give the filtered probability space and the Brownian motion martingale, in addition to the process X, to be used in (10.4). For brevity, we shall sometimes say X is a weak solution, provided it is clear what one should use for the associated filtered probability space and Brownian motion martingale.

2. In (i) above, the condition that B is a martingale relative to $\{\mathcal{F}_t\}$ can be replaced by the equivalent condition that $B(t + \cdot) - B(t)$ is independent of $\mathcal{F}_t$ for each $t \geq 0$ (see the proof of Theorem 6.1 and Exercise 6.1).

3. Clearly, any strong solution of (10.1) is also a weak solution, but the converse may not hold. A famous example of Tanaka will be used to illustrate this later in this section.

Definition. We say *weak existence* holds for the SDE (10.1) if, for each Borel probability measure μ on $\mathbb{R}^d$, there is a weak solution of (10.1) with μ as its initial law. We say *weak uniqueness* holds for the SDE (10.1) if, for each Borel probability measure μ on $\mathbb{R}^d$, all solutions of (10.1) with μ as initial distribution have the same law, i.e., if $(\Omega, \mathcal{F}, \{\mathcal{F}_t\}, P, B, X)$ and $(\tilde{\Omega}, \tilde{\mathcal{F}}, \{\tilde{\mathcal{F}}_t\}, \tilde{P}, \tilde{B}, \tilde{X})$ are two weak solutions of (10.1), both with initial distribution μ, then

$$P(X \in A) = \tilde{P}(\tilde{X} \in A) \quad \text{for all } A \in \mathcal{M},$$

where $\mathcal{M} = \sigma\{w(s) : 0 \leq s < \infty\}$ is the natural σ-field on $C_{\mathbb{R}^d}$. Weak uniqueness is also commonly referred to as uniqueness in law, or distributional uniqueness.

Example. By the last sentence of Theorem 10.6, under conditions (10.25)–

(10.28) on σ and b, any two weak solutions of (10.4), possibly on different filtered probability spaces but with the same initial distribution, are equivalent in law. Hence weak uniqueness holds for (10.1) under these conditions.

There is no obvious relationship between strong uniqueness and weak uniqueness. However, there is a third notion of uniqueness for (10.1) known as *pathwise uniqueness* which implies both strong and weak uniqueness.

Definition. We say *pathwise uniqueness* holds for the SDE (10.1) if, whenever X and $\tilde{X}$ are two weak solutions of (10.1) on the same filtered probability space with the same Brownian motion and the same initial random variable, we have

$$P(X(t) = \tilde{X}(t) \quad \text{for all } t \geq 0) = 1,$$

i.e., X and $\tilde{X}$ are indistinguishable.

Remark. There is a stronger notion of pathwise uniqueness called *strict pathwise uniqueness* where the filtrations for X and $\tilde{X}$ are allowed to be different. It turns out that pathwise uniqueness implies strict pathwise uniqueness. The proof of this and that pathwise uniqueness implies weak uniqueness uses a clever trick of Yamada and Watanabe [82] of transferring any two solutions (possibly on different probability spaces) to a common space with the same Brownian motion and the same filtration. This device is at the core of Yamada and Watanabe's proof of the following result.

Theorem 10.13. *Weak existence plus pathwise uniqueness implies strong existence and weak uniqueness.*

Remark. It is trivial that pathwise uniqueness implies strong uniqueness, as we have defined it.

We shall not prove this theorem since the proof is beyond the scope of this book. However the interested reader is referred to Ikeda and Watanabe [45, pp. 149–152], Rogers and Williams [68, §V.17] or Revuz and Yor [64]. In consulting these references, the reader should note that the theorems in [45, 68] give *equivalences* between weak existence and pathwise uniqueness on the one hand and strong existence and strong uniqueness on the other. We have only stated the result in one direction here because this is the way one usually applies the theorem and also because our notion of strong solution is slightly weaker than that in [45, 68], as discussed at the beginning of this section. We also note in passing that Yamada and Watanabe's method of proof is quite general in that it can often be applied to stochastic differential equations other than those of the form (10.1), such as those describing *reflected* diffusion processes where dt in the term $b(X(t))dt$ is replaced by the differential of a process that is locally of bounded variation.

Example. From Lemma 10.7 we see that pathwise uniqueness holds when σ and b satisfy the Lipschitz and growth conditions (10.25)–(10.28). By Theorem 10.13, this implies weak uniqueness for (10.1), a fact that was already established by direct means in Theorems 10.4 and 10.6.

The following famous example is originally due to Tanaka. It shows that

(i)   one can have a weak solution without having a strong solution, and

(ii)   one can have weak uniqueness without having pathwise uniqueness.

Example. *(Tanaka's example.)* Consider the stochastic differential equation:

$$(10.48) \qquad dX(t) = \text{sgn}(X(t))dB(t),$$

where $\text{sgn}(x)$ is 1 or -1 as x is > 0 or ≤ 0, respectively. From Section 7.3 we know that this equation has a weak solution. For if X is a one-dimensional Brownian motion martingale defined on some filtered probability space $(\Omega, \mathcal{F}, \{\mathcal{F}_t\}, P)$ then

$$B(t) = \int_0^t \text{sgn}(X(s))\, dX(s)$$

is a Brownian motion, being a continuous local martingale (relative to $\{\mathcal{F}_t\}$)

with quadratic variation t. Now, since $(\text{sgn}(x))^2 = 1$, we have

(10.49)
$$
\begin{aligned}
X(t) &= X(0) + \int_0^t \text{sgn}(X(s))\,\text{sgn}(X(s))\,dX(s) \\
&= X(0) + \int_0^t \text{sgn}(X(s))\,dB(s),
\end{aligned}
$$

where we have used the substitution theorem 2.12 for the last equality. Thus $(\Omega, \mathcal{F}, \{\mathcal{F}_t\}, P, B, X)$ is a weak solution of (10.48). Note that since X spends zero Lebesgue time at the origin, by the L^2-isometry for stochastic integrals we have $\int_0^t 1_{\{0\}}(X(s))\,dB(s) = 0$ P-a.s. It follows that it is irrelevant how $\text{sgn}(x)$ is defined at $x = 0$ for the purposes of evaluating the stochastic integral in (10.49). Consequently, on multiplying (10.49) by -1 we obtain

$$
-X(t) = -X(0) + \int_0^t \text{sgn}(-X(s))\,dB(s),
$$

and so $(\Omega, \mathcal{F}, \{\mathcal{F}_t\}, P, B, -X)$ is also a weak solution of (10.48). In particular, if $X(0) = 0$ then X and $-X$ are two distinct weak solutions of (10.48) on the same filtered probability space with the same initial value. Thus pathwise uniqueness cannot hold for (10.48).

On the other hand, for any weak solution $(\Omega, \mathcal{F}, \{\mathcal{F}_t\}, P, B, X)$ of (10.48), X is a continuous local martingale with quadratic variation (see Corollary 5.9):

$$
[X]_t = \int_0^t \{\text{sgn}(X(s))\}^2\,ds = t \quad \text{for all } t \geq 0.
$$

Thus, by Theorem 6.1, any weak solution of (10.48) is a Brownian motion and so, given its initial distribution, it is unique in law. Hence, weak uniqueness holds for (10.48). In addition, by Tanaka's formula (7.15) applied to such an X we have

(10.50)
$$
|X(t)| - |X(0)| - L_X(t, 0) = \int_0^t \text{sgn}(X(s))\,dX(s),
$$

where $L_X(\,\cdot\,, 0)$ is the local time of X at the origin. By an argument similar to that leading to (10.49), but with dB in place of dX, we have

$$B(t) = \int_0^t \operatorname{sgn}(X(s))\operatorname{sgn}(X(s))\,dB(s)$$

(10.51)

$$= \int_0^t \operatorname{sgn}(X(s))\,dX(s).$$

By combining (10.51) with (10.50) and the fact that $L_X(\,\cdot\,,0)$ is adapted to the standard filtration generated by $|X|$ (see (7.1)), we conclude that B is adapted to the standard filtration generated by $|X|$. Thus X cannot be adapted to the standard filtration generated by $X(0)$ and B. It follows that there cannot be a strong solution of (10.48) for any quintuple of data $(\Omega, \mathcal{F}, P, B, \xi)$.

An often convenient means for proving weak existence and uniqueness is via martingale problems, as developed by Stroock and Varadhan [73]. This approach lies somewhere between the direct stochastic differential equation approach of Section 10.2 and a potential theoretic or semigroup approach. A major advantage of this formulation is that one can often prove weak existence by establishing existence for an approximating family of processes and then passing to a weak limit. Moreover, when there is uniqueness for the martingale problem, one can usually establish the strong Markov property for the family of probability measures solving the martingale problem.

For the formal definition of the martingale problem associated with σ and b, let $a = \sigma\sigma'$, where $'$ denotes transpose, and define

$$Lf = \frac{1}{2}\sum_{i,j=1}^d a_{ij}\frac{\partial^2 f}{\partial x_i \partial x_j} + \sum_{i=1}^d b_i\frac{\partial}{\partial x_i} \quad \text{for } f \in C^2(\mathbb{R}^d).$$

Let $C_b^2(\mathbb{R}^d)$ denote the space of twice continuously differentiable functions on $\mathbb{R}^d$ that together with their first and second derivatives are bounded there.

Definition. A solution of the *martingale problem* associated with σ and b is a family of probability measures $\{P^x,\ x \in \mathbb{R}^d\}$ on $(C_{\mathbb{R}^d}, \mathcal{M})$ such that for each $x \in \mathbb{R}^d$,

(i) $P^x(w(0) = x) = 1,$

(ii)   for each $f \in C_b^2(\mathbb{R}^d)$,

$$\left\{ f(w(t)) - \int_0^t (Lf)(w(s))\, ds,\, \mathcal{M}_t,\, t \geq 0 \right\}$$

is a martingale on $(C_{\mathbb{R}^d}, \mathcal{M}, P^x)$.

(Note here that contrary to our usual convention, the filtration $\{\mathcal{M}_t\}$ has not been completed.)

It can be shown that there is a unique weak solution of (10.1) for each initial Borel probability measure μ on $\mathbb{R}^d$, if and only if there is a unique solution of the martingale problem associated with σ and b. Indeed, it is a simple exercise using Itô's formula to show that any weak solution of (10.1) with starting point x generates a solution of (i) and (ii) above (see Exercise 4). The converse is more difficult. One first observes that for a unique solution $\{P^x, x \in \mathbb{R}^d\}$ of the martingale problem, by integrating P^x with respect to $\mu(dx)$, one can obtain a law P^μ under which $w(\cdot)$ has initial law μ. (The measurability of $x \to P^x$ required for this comes from the uniqueness of the martingale problem (see [73, p. 152]).) Then, one can show that $w(\cdot)$ under P^μ is a weak solution of (10.1). For this, when σ does not have a bounded inverse, one may have to expand the probability space, introducing some Brownian components independent of w. We shall not go into the details here, but refer the interested reader to Karatzas and Shreve [49, p. 314]. It can be shown that when a and b are bounded and a is uniformly continuous and uniformly positive definite, there is a unique solution of the martingale problem (see [73, p. 187]).

Remark. For $d = 1$, recent results of Engelbert and Schmidt [29, 30] give necessary and sufficient conditions for weak existence and uniqueness for the SDE (10.1) with $b = 0$ and sufficient conditions in the case $b \neq 0$. The reader is referred to Karatzas and Shreve [49, §5.5] or [29, 30] for details.

10.5 Examples.

Example. *(Bessel Process.)* Consider the stochastic differential equation

$$(10.52) \qquad\qquad dX_t = \frac{\alpha}{2} X_t^{-1} \, dt + dB_t,$$

where B is a one-dimensional Brownian motion and $\alpha \in I\!\!R$. For $\alpha = 0$, (10.52) obviously has a strong solution for any given initial value in $I\!\!R$, but for $\alpha \neq 0$, the drift is singular at the origin and some care is required to interpret what the equation means there.

By the Bessel process example in Section 5.4, we know that for $\alpha = n-1$, $n \in \{2, 3, \dots\}$, and $x \in (0, \infty)$, there is a solution of (10.52) with initial value x and paths in $(0, \infty)$. These solutions are actually strong solutions. This fact can be deduced from the analysis below, or by considering $Y = X^2$ and invoking the results on strong solutions for Hölder coefficients when $d = 1$ (see Exercise 5). We shall now investigate how one can apply the results of Theorem 10.5 locally to obtain strong solutions of (10.52) for all $\alpha \geq 1$ and starting points $x \in (0, \infty)$. By symmetry this is equivalent to considering $\alpha \leq -1$ for $x \in (-\infty, 0)$. At the outset, we shall not restrict the values of α, except to exclude the trivial case $\alpha = 0$. We suppose $x \in (0, \infty)$.

In the following, the term adapted will refer to the standard filtration generated by a given one-dimensional Brownian motion B that starts from the origin. Consider $I = [a, c]$ for $0 < a \leq x \leq c < \infty$, and let b^I be a bounded uniformly Lipschitz function on $I\!\!R$ such that $b^I(y) = \alpha/2y$ for $y \in I$. Then b^I and $\sigma \equiv 1$ satisfy conditions (10.6)–(10.7), and it follows from Theorem 10.5 that there is a unique continuous adapted process X^I that satisfies

$$dX_t^I = b^I(X_t^I)dt + dB_t \quad \text{for all } t \geq 0,$$

with $X_0^I = x$. Moreover, it follows as in Lemma 10.7 that for $\tau^I = \inf\{t \geq 0 : X_t^I \notin I\}$, $X^I(\cdot \wedge \tau^I)$ is uniquely determined, regardless of the values of b^I outside of I. As in the proof of Theorem 10.6, if we can show that $\tau^I \uparrow \infty$ a.s. as $a \downarrow 0$ and $c \uparrow \infty$, it will follow that $X(\cdot) = \lim_{c \uparrow \infty} \lim_{a \downarrow 0} X^I(\cdot \wedge \tau^I)$ is the unique continuous adapted solution of (10.52) with initial condition

$X_0 = x \in (0, \infty)$ and

$$P(X_t > 0 \quad \text{for all } t \ge 0) = 1.$$

Conditions for $\tau^I \uparrow \infty$ as $a \downarrow 0$ and $c \uparrow \infty$ can be found using the techniques illustrated in the exercises of Chapter 6. We only sketch the details in (i)–(iii) below, leaving the full details to the reader as an exercise (see Exercise 6).

(i) We first show that $E[\tau^I] < \infty$ for each I. Let $f(y) = y$ for $y \in \mathbb{R}$. Applying Itô's formula to X^I and f, after stopping at τ^I and taking expectations, one obtains

$$E\left[f(X^I(t \wedge \tau^I)) - f(X^I(0))\right] = \frac{\alpha}{2} E\left[\int_0^{t \wedge \tau^I} \{X^I(s)\}^{-1} \, ds\right] \quad \text{for all } t \ge 0.$$

Multiplying by the non-zero quantity $2\alpha^{-1}$ and taking an upper bound for the resulting left member and a lower bound for the resulting right member, we obtain

$$\frac{2}{|\alpha|} c \ge c^{-1} E\left[t \wedge \tau^I\right].$$

Letting $t \to \infty$, we obtain

$$E\left[\tau^I\right] \le \frac{2c^2}{|\alpha|} < \infty.$$

(ii) Now consider $g(y) = y^{1-\alpha}$ for $\alpha \ne 1$ and $g(y) = \ln y$ for $\alpha = 1$, which satisfies

$$\frac{1}{2} g''(y) + \frac{\alpha}{2y} g'(y) = 0 \quad \text{for } y \in (0, \infty).$$

Then, by Itô's formula we have

$$E\left[g(X^I(t \wedge \tau^I)) - g(X^I(0))\right] = 0 \quad \text{for all } t \ge 0.$$

Since we know $\tau^I < \infty$ a.s. from (i) above, we can let $t \to \infty$ to obtain by bounded convergence that

$$E\left[g(X^I(\tau^I))\right] = g(x).$$

Setting $\tau_y^I = \inf\{t \geq 0 : X_t^I = y\}$ for each $y \in \mathbb{R}$, the above reduces for $\alpha \neq 1$ to

$$P(\tau_a^I < \tau_c^I)a^{1-\alpha} + P(\tau_c^I < \tau_a^I)c^{1-\alpha} = x^{1-\alpha}.$$

Since $P(\tau_a^I < \tau_c^I) + P(\tau_c^I < \tau_a^I) = P(\tau^I < \infty) = 1$, the above yields

$$(10.53) \qquad P(\tau_c^I < \tau_a^I) = \frac{x^{1-\alpha} - a^{1-\alpha}}{c^{1-\alpha} - a^{1-\alpha}}, \qquad \alpha \neq 1.$$

Similarly, for $\alpha = 1$, we have

$$(10.54) \qquad P(\tau_c^I < \tau_a^I) = \frac{\ln x - \ln a}{\ln c - \ln a}.$$

It follows that for $\alpha \geq 1$, $\lim_{a \to 0} P(\tau_c^I < \tau_a^I) = 1$ and so almost surely, $\lim_{a \to 0} X^I(\tau^I) = c$. In this case, setting $\tau_c = \lim_{a \downarrow 0} \tau^I$ and defining $X^c(t) \equiv \lim_{a \downarrow 0} X^I(t \wedge \tau_I)$, we see that X^c is a continuous adapted process living on $(0, c]$ and it satisfies

$$X^c(t \wedge \tau_c) = x + \frac{\alpha}{2} \int_0^{t \wedge \tau_c} (X_s^c)^{-1} \, ds + B(t \wedge \tau_c) \quad \text{for all} \quad t \geq 0.$$

In particular, X^c does not reach the origin in finite time. Using the same technique as in the proof of Theorem 10.6 (cf. (10.32)) it can then be shown that $\tau_c \uparrow \infty$ a.s. as $c \uparrow \infty$, and it follows that $X(\cdot) \equiv \lim_{c \uparrow \infty} X^c(\cdot \wedge \tau_c)$ defines a solution of (10.52). Thus, when $\alpha \geq 1$, there is a strong solution of (10.52) starting from each $x \in (0, \infty)$.

(iii) For $\alpha < 1$, one can prove using (10.53)–(10.54) that the continuous process on $[0, \tau_c)$, $\tau_c \equiv \lim_{a \downarrow 0} \tau_c^I$, obtained by patching together the local solutions $X^I(\cdot \wedge \tau^I)$, has a positive probability of approaching the origin as $t \uparrow \tau_c < \infty$. In this case of $\alpha < 1$, and when $x = 0$ for $\alpha \geq 0$, one needs to address the problem that equation (10.52) does not make sense there. For this, one must prescribe an additional condition describing the desired behavior of the process there. One can always simply require that the process be absorbed at the origin. The feasibility of other behaviors that result in a continuous extension of the process beyond the time of reaching the origin, such as instantaneous reflection at the origin or passage through to $(-\infty, 0)$, depends on the value of α. For $\alpha > -1$, it is possible for the process to be extended continuously so as to return to $(0, \infty)$ (reflection at the origin); for $\alpha < 1$, the process may pass continuously through to $(-\infty, 0)$; and for $-1 < \alpha < 1$, a combination of these behaviors is possible (skew reflection). These results follow from the general theory of boundary behavior for one-dimensional diffusions. We refer the interested reader to [51, Chapter 15] or [47] for the justification of these claims and further discussion of this expansive topic. For recent work on extending the meaning of (10.52) to the realm $0 < \alpha < 1$ using local times, see Bertoin [2].

Example. *(Black-Scholes option pricing formula.)* In the following B is a one-dimensional Brownian motion, starting from the origin, defined on a complete probability space $(\Omega, \mathcal{F}, P)$, and $\{\mathcal{F}_t\}$ is the standard filtration generated by B (see §1.8).

We consider a financial market in which two financial securities are available, one labelled *stock* and the other *bond*. The stock is a risky asset whose price S_t per share at time t is governed by the stochastic differential equation

$$(10.55) \qquad dS_t = \mu S_t dt + \sigma S_t dB_t, \quad t \geq 0,$$

where $\mu \neq 0$, $\sigma > 0$ and $S_0 > 0$ are constants. The purely formal notation

$$\frac{dS_t}{S_t} = \mu dt + \sigma dB_t$$

suggests that μ may be interpreted as the mean rate of return for the stock and σ^2 may be interpreted as the variance of the rate of return. Observe that the coefficients in (10.55) satisfy conditions (10.6)–(10.7), and consequently there is a unique strong solution of (10.55) with initial value S_0. Indeed, one can readily verify using Itô's formula that

$$(10.56) \qquad S_t = S_0 \exp\left(\left(\mu - \frac{1}{2}\sigma^2\right)t + \sigma B_t\right), \quad t \geq 0,$$

satisfies (10.55) (cf. (6.7)). The bond is a riskless asset whose price β_t per share at time t is given by $\beta_t = \beta_0 e^{rt}$, where $r > 0$ is the interest rate and β_0 is a positive constant.

If an investor purchases a shares of stock at time t and sells them at a later time $t + h$ ($h > 0$), then the realized capital gain is $a(S_{t+h} - S_t)$.

More generally, an investor could adjust his stock holdings over a fixed time period $[0, T]$ by holding a constant amount over each time interval in some partition $(t_0, t_1], (t_1, t_2], \ldots, (t_{n-1}, t_n]$ of $(0, T]$, where $t_0 = 0$, $t_n = T$. The process representing the amount of stock held at any time t is then given by

$$a_t = \sum_{j=1}^{n} a_{t_j} 1_{(t_{j-1}, t_j]}(t), \quad t \in [0, T],$$

and the realized capital gain over $[0, T]$ is then

$$\sum_{j=1}^{n} a_{t_j}(S_{t_j} - S_{t_{j-1}}).$$

But the latter is simply the stochastic integral of a with respect to dS over $[0, T]$, defined by

$$(10.57) \qquad \int_0^T a_t dS_t \equiv \mu \int_0^T a_t S_t dt + \sigma \int_0^T a_t S_t dB_t.$$

Furthermore, assuming a non-clairvoyant investor, it makes sense to allow the a_{t_j} to be random variables that depend on the information available up to the beginning of the investment period $(t_{j-1}, t_j]$, i.e., to allow $a_{t_j} \in \mathcal{F}_{t_{j-1}}$ for $j = 1, \ldots, n$. Then $\{a_t, t \in [0, T]\}$ is a predictable process and the capital gain over $[0, T]$ is still given by (10.57). Idealizing the scheme one can model "continuous trading" in the stock by allowing a to be any predictable process with sufficient integrability that the stochastic integral $\int_0^t a_s dS_s$, defined by (10.57) (with t there in place of T), is well defined for all $t \in [0, T]$ (see §2.6). Then $\int_0^T a_t dS_t$ represents the capital gain over $[0, T]$ from such an investment strategy. (In fact, since the stochastic integrator in (10.57) is a Brownian motion, one could even further relax the measurability requirements on a, as described in Chapter 3. However, for our purposes it suffices to consider predictable integrands; indeed, continuous adapted integrands will suffice.) Similarly, one can incorporate continuous trading in the bond by allowing $b = \{b_t, t \in [0, T]\}$ to be any predictable process such that $\int_0^t b_s d\beta_s$, $t \in [0, T]$ defines a continuous adapted process. Then, b represents a strategy of holding b_t shares of the bond at time t, for each $t \in [0, T]$, and $\int_0^T b_t d\beta_t$ represents the realized capital gain over $[0, T]$ from such an investment. A *trading strategy* is a pair (a, b) as described above. Such a strategy is said to be *self-financing* if P-a.s.,

$$(10.58) \quad a_t S_t + b_t \beta_t = a_0 S_0 + b_0 \beta_0 + \int_0^t a_s \, dS_s + \int_0^t b_s d\beta_s \quad \text{for all } t \in [0, T].$$

The interpretation of (10.58) is that the current portfolio value V_t, given by the left member of (10.58), is precisely equal to the inital investment plus capital gains to date. We shall allow a_t and b_t to take negative values as well as positive ones, but the total portfolio value V_t should be non-negative. Negative values of a_t and b_t correspond to "short sales". In particular, a negative value for a_t represents a short sale of stock by the investor, i.e., he sells the stock at time t, but does not have to deliver the stock until time T. Negative values of b_t correspond to borrowing at the bond's riskless interest rate r.

Suppose it is possible to purchase a ticket at time zero that enables the holder to buy one share of stock at time T at a fixed price K, called the *exercise price*. Such a ticket is called a *European call option*, in contrast to an *American call option* which can be exercised at any time in the interval $[0, T]$. If at time T the stock price S_T is less than or equal to K, the option will not be exercised and the ticket expires as a worthless contract. On the other hand, if the stock price S_T is greater than K, the ticket holder can buy one share of stock at price K and then turn around and sell it at the price S_T for a net profit of $S_T - K$. Thus, the option is equivalent to a ticket which entitles the bearer to a payment of $(S_T - K)^+$ at time T.

A basic question concerning this financial model is "How much would you be willing to pay for such a ticket at time zero?" Put another way, "What is a 'rational value' to pay for the option at time zero?" In their famous paper [4], Black and Scholes derived a formula for the rational value; it is the unique amount of money such that

(i)   an individual, after investing this amount of money in stock and bond at time zero, can manage his portfolio according to a self-financing strategy so as to yield the same payoff $(S_T - K)^+$ as if the option had been purchased, and

(ii)   if the option were offered at any price other than this amount, there would be the opportunity for arbitrage, i.e., for unbounded profits without an accompanying risk of loss.

In fact the rational value is characterized by property (i) alone. A proof of this was given in Harrison and Pliska [40], which built on the earlier work of Harrison and Kreps [38] and Merton [58]. Following the approach outlined in Section 1.2 of [40], we will indicate below how one might guess the rational value and identify an associated self-financing strategy. This is followed by a discussion of the rationale based on arbitrage, which makes use of the existence of a self-financing strategy as described in (i).

Suppose we try to find a self-financing strategy (a, b) with associated value process V such that

$$(10.59) \qquad V_t = f(S_t, T - t), \quad \text{for all } t \in [0, T],$$

for some function $f \in C^{2,1}((0, \infty) \times (0, T]) \cap C((0, \infty) \times [0, T])$ and $f(x, 0) = (x - K)^+$. It is natural to consider such a form for the value process if one is seeking a value process that has a Markovian dependence on S. Here $C^{2,1}((0, \infty) \times (0, T])$ denotes the space of real-valued functions $f(x, s)$, that are twice continuously differentiable in $x \in (0, \infty)$ and once continuously differentiable in $s \in (0, T]$. Note that since S is a positive process we only need the x variable of f to range over $(0, \infty)$. Now for any f as described above and V defined by (10.59), writing partial derivatives of f with respect to x as f_x, f_{xx}, and with respect to s as f_s, by Itô's formula we have for all $0 \leq t < T$,

$$(10.60) \qquad \begin{aligned} V_t - V_0 &= \int_0^t f_x(S_u, T - u) dS_u \\ &- \int_0^t f_s(S_u, T - u) du + \frac{1}{2} \int_0^t \sigma^2 S_u^2 f_{xx}(S_u, T - u) du. \end{aligned}$$

On the other hand, V will be the value corresponding to a self-financing strategy (a, b) if and only if for all $t \in [0, T]$,

$$(10.61) \qquad V_t - V_0 = \int_0^t a_u dS_u + \int_0^t b_u d\beta_u \quad \text{and} \quad b_t = (V_t - a_t S_t)/\beta_t.$$

Comparing (10.60) and (10.61), and recalling that $\beta_t = \beta_0 e^{rt}$, we see that (10.61) holds for all $t \in [0, T)$ (we will deal with the case $t = T$ later), if for all such t,

$$(10.62) \qquad a_t = f_x(S_t, T - t)$$

and

$$r(f(S_t, T - t) - S_t f_x(S_t, T - t)) = -f_s(S_t, T - t) + \frac{1}{2}\sigma^2 S_t^2 f_{xx}(S_t, T - t).$$

The last equation above will hold if f satisfies

$$(10.63) \qquad f_s = \frac{1}{2}\sigma^2 x^2 f_{xx} + rx f_x - rf \quad \text{for } (x, s) \in (0, \infty) \times (0, T].$$

Moreover, to ensure that $V_T = (S_T - K)^+$, we want

$$(10.64) \qquad f(x, 0) = (x - K)^+, \quad x \in (0, \infty).$$

Now, one can try to directly solve the parabolic equation (10.63) with (10.64) as initial condition, or alternatively, one can obtain a probabilistic representation for such a solution, and by evaluating this probabilistic expression obtain a candidate for a solution of (10.63)–(10.64). We follow the latter procedure below, where we use a change of measure transformation. Harrison and Kreps [38] were the first to recognize the utility of such a transformation in this setting.

Let $\tilde{P}$ be the measure on $\mathcal{F}_T$ that is equivalent to P and whose Radon-Nikodym derivative is given by

$$\frac{d\tilde{P}}{dP} = \exp\left(\frac{(r - \mu)}{\sigma}B_T - \frac{1}{2}\left(\frac{r - \mu}{\sigma}\right)^2 T\right).$$

By the results of Section 9.4 on change of measure, for

$$\tilde{B}_t \equiv B_t + \left(\frac{\mu - r}{\sigma}\right)t,$$

$\{\tilde{B}_t, t \in [0, T]\}$ is a Brownian motion on $[0, T]$ under $\tilde{P}$. Equation (10.55) implies that we have $\tilde{P}$-a.s.,

$$(10.65) \qquad S_t = S_0 + r\int_0^t S_u \, du + \sigma\int_0^t S_u \, d\tilde{B}_u, \quad \text{for all } t \in [0, T].$$

Observe from (10.56) that $\tilde{P}$-a.s.,

$$S_t = S_0 \exp\left(\left(r - \frac{1}{2}\sigma^2\right)t + \sigma\tilde{B}_t\right), \quad t \in [0, T],$$

and then by (9.26),

$$(10.66) \qquad \tilde{E}[S_t^2] = S_0^2 \tilde{E}[\exp((2r - \sigma^2)t + 2\sigma\tilde{B}_t)] = S_0^2 \exp(2rt + \sigma^2 t),$$

where $\tilde{E}$ denotes expectation with respect to $\tilde{P}$. Now for $f \in C^{2,1}((0, \infty) \times (0, T])$, by Itô's formula and (10.65) we have $\tilde{P}$-a.s. for all $t \in [0, T)$,

$$
\begin{aligned}
e^{-rt} &f(S_t, T - t) - f(S_0, T) \\
&= \sigma \int_0^t e^{-ru} f_x(S_u, T - u) S_u d\tilde{B}_u + r \int_0^t e^{-ru} f_x(S_u, T - u) S_u du \\
&\quad - \int_0^t e^{-ru} f_s(S_u, T - u) du - r \int_0^t e^{-ru} f(S_u, T - u) du \\
&\quad + \frac{1}{2}\sigma^2 \int_0^t e^{-ru} f_{xx}(S_u, T - u) S_u^2 du.
\end{aligned}
$$

Now, if f satisfies (10.63), the above reduces to

$$(10.67) \qquad e^{-rt} f(S_t, T - t) - f(S_0, T) = \sigma \int_0^t e^{-ru} f_x(S_u, T - u) S_u d\tilde{B}_u.$$

Now from (10.66) it follows that $\{S_t, t \in [0, T)\}$ is L^2-bounded, and hence by the basic L^2-isometry for stochastic integrals, if f_x is bounded, then the stochastic integral with respect to $d\tilde{B}_u$ above is an L^2-bounded $\tilde{P}$-martingale on $[0, T)$. It follows that the stochastic integral can be defined for $t = T$ as its L^2 limit as $t \uparrow T$ and the resulting stochastic integral process is an L^2-martingale on $[0, T]$ (cf. Theorem 1.5). Furthermore, if $f \in C((0, \infty) \times [0, T])$, we can let $t \uparrow T$ in the left member of (10.67), so that under the above conditions, (10.67) holds for all $t \in [0, T]$. Then on taking expectations in this equation with $t = T$ and supposing that f satisfies (10.64), we obtain

$$(10.68) \qquad f(S_0, T) = \tilde{E}\left[e^{-rT} f(S_T, 0)\right] = \tilde{E}\left[e^{-rT}(S_T - K)^+\right].$$

Now, using the explicit exponential formula for S_T in terms of the $\tilde{P}$-Brownian motion $\tilde{B}$, the right member of (10.68) can be evaluated after a somewhat tedious calculation. Replacing S_0 by x and T by t, we obtain the candidate function

(10.69) $$f(x,t) = x\Phi(g(x,t)) - Ke^{-rt}\Phi(h(x,t)),$$

where

$$g(x,t) = \left[\ln(x/K) + \left(r + \frac{1}{2}\sigma^2\right)t\right] \Big/ \sigma\sqrt{t},$$

$$h(x,t) = g(x,t) - \sigma\sqrt{t},$$

and Φ is the standard normal distribution function. One can now verify that f given by (10.69) (with the obvious limit value at $t = 0$), is such that $f \in C^{2,1}((0,\infty) \times (0,T]) \cap C((0,\infty) \times [0,T])$, $f_x = \Phi(g(x,t))$ is bounded for $(x,t) \in (0,\infty) \times (0,T]$, and f satisfies (10.63)–(10.64). Since $\tilde{P}(S_T = K) = 0$, for a_t defined by (10.62) for $t \in [0,T)$, we have $\lim_{t\uparrow T} a_t = 1_{\{S_T > K\}}$ $\tilde{P}$-a.s. Thus, by defining a_T to equal this limiting value, we see that $\{a_t, t \in [0,T]\}$ is a $\tilde{P}$-a.s. continuous adapted process. Define $b_t = (f(S_t, T-t) - a_t S_t)/\beta_t$, for all $t \in [0,T]$. It then follows from the above analysis that $f(S_0, T) = \tilde{E}[e^{-rT}(S_T - K)^+]$ is an amount that can be invested according to the self-financing strategy (a,b) to yield a net portfolio value at time T of $(S_T - K)^+$. It is clear from the representation (10.68) and (10.65) that the value $f(S_0, T)$ is independent of μ. By direct computation one can verify that $a_t \geq 0$ for all $t \in [0,T]$, although b_t may take both positive and negative values over the time interval $[0,T]$. That is, the self-financing strategy (a,b) may require borrowing money at the bond's interest rate r, but it does not require short sales of stock.

To understand the interpretation of $q \equiv f(S_0, T)$ as a rational value in terms of arbitrage, suppose the initial option price p is something other than q. If $p > q$, consider the following strategy. At time $t = 0$,

(i)   sell the option to someone else at the price p, and

(ii)   invest an amount q in stock and bond according to the self-financing strategy (a,b).

This strategy nets an initial profit of $p - q > 0$. At time T, the terminal portfolio value is $a_T S_T + b_T \beta_T = (S_T - K)^+$ and the investor is obligated to pay $(S_T - K)^+$ in order to meet his obligation to the purchaser of the option. That is, if $S_T > K$, he must buy the stock for S_T and sell it to the option holder at the exercise price K, for a net loss of $S_T - K$; or if $S_T \leq K$, he has no obligation, since the option will not be exercised. Thus, the total terminal profit is zero and hence the net profit is $p - q$. The scale of this strategy can be increased by selling n options and investing nq in stock and bond according to the self-financing strategy (na, nb) for a net profit of $n(p - q)$. Thus, if $p > q$, there is the opportunity for arbitrarily large profits without an accompanying risk of loss, i.e., for arbitrage. On the other hand, if $p < q$, then an investor can implement an arbitrage strategy by buying options and selling stock and bond. For suppose he buys the option at time zero and pursues the strategy $(-a, -b)$, i.e., he sells when he would buy according to (a, b) and vice versa. We note that this will involve short sales of stock, and may involve borrowing of money. The investor realizes an initial profit of $q - p$, and at time T he must produce stock and bond with a total value of $(S_T - K)^+$. But this can be achieved by exercising the option if $S_T > K$, and not exercising it otherwise. Hence the terminal profit is zero and the net profit is $q - p > 0$. Again, the scale can be increased to yield arbitrarily large profits with no accompanying risk of loss. Thus, the only price for the option that could possibly exclude arbitrage is $q = f(S_0, T)$.

Alert readers will wonder whether, in view of the explicit representation (10.68), there is not a more probabilistic method for showing that the final payoff $(S_T - K)^+$ is attainable by a self-financing strategy from an initial investment amount given by the right member of (10.68). Indeed, there is such a derivation, but it requires the use of a powerful theorem on representation of martingales adapted to a Brownian motion, a topic that we have not covered in this book; see Exercise 7 below. The approach we adopted here is more elementary and self-contained. In addition, it has allowed us to illustrate one of the many connections between Itô's formula and partial differential equations. For generalizations of the Black-Scholes formula and discussion of related topics, we refer the interested reader to Duffie [26].

10.6 Exercises.

1. Prove that Lemma 10.1 still holds when σ and b are not bounded, but still satisfy (10.6)–(10.7), and Y and Z are continuous.

2. Prove Gronwall's inequality, Lemma 10.2.
Hint: Let $v(t) = \int_0^t f(s)ds$ and $w(t) = \int_0^t v(s)ds$, so that $v(t) = \frac{dw}{dt}$ for all $t \in [0, T]$. Integrate the given inequality and write it in terms of $w(t)$ and its derivative. Multiplying by an integrating factor and then integrating yields an inequality for $w(t)$ which when back-substituted yields an inequality for $v(t)$ and hence for $f(t)$.

3. Verify that the process X, defined in the existence part of the proof of Theorem 10.6, does in fact satisfy (10.4).

4. Suppose X^x is a weak solution of (10.1) with initial condition $X_0 = x$. Prove that the probability measure induced on $(C_{\mathbb{R}^d}, \mathcal{M})$ by X^x satisfies conditions (i) and (ii) of the martingale problem in Section 10.4.
Hint: First apply Itô's formula to X^x on its original probability space. Then transfer the resulting martingale property of $f(X_t^x) - \int_0^t Lf(X_s^x)\, ds$ to the canonical path space using the fact that $\{g_1(w(t_1)), \ldots, g_n(w(t_n)); 0 \le t_1 < \ldots < t_n \le t; g_1, \ldots, g_n \in C_b(\mathbb{R}^d); n \in \mathbb{N}\}$ generates $\mathcal{M}_t$.

5. Assuming the result concerning Hölder coefficients described at the end of Section 10.3, show that for each $\alpha \in (1, \infty)$ any weak solution of (10.52) with starting point $x \in (0, \infty)$ is also a strong solution of that equation. Hence, in the Bessel process example of Section 5.4, R is actually adapted to the filtration generated by the Brownian motion W.
Hint: Consider the SDE for $Y = X^2$ and note that X can be recovered from Y.

6. Provide full details of (i) and (ii) in the proof for the Bessel process that $\tau_I \uparrow \infty$ as $a \downarrow 0$ and $c \uparrow \infty$.

7. Assume the setup of the Black-Scholes option pricing example in Section 10.5. In particular, B is a one-dimensional Brownian motion and $\{\mathcal{F}_t, t \ge 0\}$ is the standard filtration generated by B. Fix $T > 0$. You may assume the following theorem. For a proof, see Rogers and Williams [68, §IV.36].

Theorem. *Suppose $Y \in L^2(\Omega, \mathcal{F}_T, P)$. Then there is a predictable process $X \in L^2([0, T] \times \Omega, \mathcal{P}, \lambda \times P)$ such that*

$$Y = E[Y] + \int_0^T X_s \, dB_s.$$

Apply the theorem with $Y = (S_T - K)^+$. For the resulting X, define $a_t = X_t/(\sigma S_t)$, $t \in [0, T]$. By solving an integral equation, find $b = \{b_t, t \in [0, T]\}$ such that (a, b) is a self-financing strategy that achieves the final payoff $(S_T - K)^+$.

8. Assuming $\mu = r = 0$, verify by explicit calculation that the right member of (10.68) is given by (10.69).

REFERENCES

[1]   Aizenman, M., and Simon, B., "Brownian motion and Harnack inequality for Schrödinger operators", *Comm. Pure and Appl. Math.*, **35** (1982), 209–273.

[2]   Bertoin, J., "Excursions of a $\mathrm{BES}_0(d)$ and its drift term ($0 < d < 1$)", *Prob. Theor. Rel. Fields*, to appear.

[3]   Billingsley, P., *Convergence of Probability Measures*, John Wiley and Sons, New York, 1968.

[4]   Black, F., and Scholes, M., "The pricing of options and corporate liabilities", *J. Polit. Econom.*, **81** (1973), 637–659.

[5]   Blumenthal, R. M., and Getoor, R. K., *Markov Processes and Potential Theory*, Academic Press, New York, 1968.

[6]   Boukricha, A., Hansen, W., and Hueber, H., "Continuous solutions of the generalized Schrödinger equation and perturbation of harmonic spaces", *Expositiones Mathematica*, **5** (1987), 97–135.

[7]   Breiman, L., *Probability*, Addison-Wesley, Reading, MA, 1968.

[8]   Cameron, R. H., and Martin, W. T., "Transformations of Wiener integrals under translations", *Ann. Math.*, **45** (1944), 386–396.

[9]   Cameron, R. H., and Martin, W. T., "Transformations of Wiener integrals under a general class of linear transformations", *Trans. Amer. Math. Soc.*, **58** (1945), 184–219.

[10]   Cameron, R. H., and Martin, W. T., "The transformation of Wiener integrals by nonlinear transformations", *Trans. Amer. Math. Soc.*, **66** (1949), 253–283.

[11]   Chung, K. L., *A Course in Probability Theory*, 2nd. ed., Academic Press, New York, 1974.

[12]   Chung, K. L., *Lectures from Markov Processes to Brownian Motion*, Springer-Verlag, New York, 1982.

[13]   Chung, K. L., "Probability methods in potential theory", *Potential Theory, Surveys and Problems, Proceedings, Prague 1987*, Lecture Notes in Mathematics, Vol. 1344, Springer-Verlag, 1988, 42–54.

[14]   Chung, K. L., and Rao, K. M., "General gauge theorem for multiplicative functionals", *Trans. Amer. Math. Soc.*, **306** (1988), 819–836.

[15]   Chung, K. L., "Reminiscences of some of Paul Lévy's ideas in Brownian motion and in Markov chains", *Seminar on Stochastic Processes 1988*, eds. E. Çinlar, K. L. Chung, R. K. Getoor, Birkhäuser, Boston, 1989, 99–107.

[16]   Chung, K. L., and Hsu, P., "Gauge theorem for the Neumann problem", *Seminar on Stochastic Processes 1984*, eds. E. Çinlar, K. L. Chung, R. K. Getoor, Birkhäuser, Boston, 1986, 63–70.

[17]   Chung, K. L., Li, P., and Williams, R. J., "Comparison of probability and classical methods for the Schrödinger equation", *Expositiones Mathematica*, **4** (1986), 271–278.

[18]   Chung, K. L., and Rao, K. M., "Feynman-Kac functional and the Schrödinger equation", *Seminar on Stochastic Processes 1981*, eds. E. Çinlar, K. L. Chung, R. K. Getoor, Birkhäuser, Boston, 1981, 1–29.

[19] Chung, K. L., and Walsh, J. B., "Meyer's theorem on predictability", *Z. Wahr. verw. Geb.*, **29** (1974), 253–256.

[20] Chung, K. L., and Zhao, Z., forthcoming book.

[21] Coddington, E. A., *An Introduction to Ordinary Differential Equations*, Prentice Hall, New Jersey, 1961.

[22] Cranston, M., Fabes, E., and Zhao, Z., "Conditional gauge and potential theory for the Schrödinger operator", *Trans. Amer. Math. Soc.*, **307** (1988), 171–194.

[23] Dellacherie, C., and Meyer, P. A., *Probabilities and Potential*, Vol. I, North-Holland, Amsterdam, 1978.

[24] Dellacherie, C., and Meyer, P. A., *Probabilités et Potentiel*, Vol. II, *Théorie des Martingales*, Hermann, Paris, 1980.

[25] Doléans-Dade, C., "Existence du processus croissant naturel associé à un potentiel de class (D)", *Z. Wahr. verw. Geb.*, **9** (1968), 309–314.

[26] Duffie, D., *Security Markets—Stochastic Models*, Academic Press, San Diego, CA, 1988.

[27] Durrett, R., *Brownian Motion and Martingales in Analysis*, Wadsworth, Belmont, CA, 1984.

[28] El Karoui, N., and Chaleyat-Maurel, M., "Un problème de réflexion et ses applications au temps local et aux équations différentielles stochastiques sur $I\!R$—Cas continu", *Astérisque*, Société Mathématique de France, **52–53** (1978), 117–144.

[29] Engelbert, H. J., and Schmidt, W., "On one-dimensional stochastic differential equations with generalized drift", *Lecture Notes in Control and Information Sciences*, Vol. 69, Springer-Verlag, New York, 1984, 143–155.

[30] Engelbert, H. J., and Schmidt, W., "On solutions of stochastic differential equations without drift", *Z. Wahr. verw. Geb.*, **68** (1985), 287–317.

[31]   Falkner, N., "Feynman-Kac functionals and positive solutions of $\frac{1}{2}\Delta u + qu = 0$", *Z. Wahr. verw. Geb.*, **65** (1983), 19–34.

[32]   Falkner, N., "Conditional Brownian motion in rapidly exhaustible domains", *Ann. Probab.*, **15** (1987), 1501–1514.

[33]   Freedman, D., *Brownian Motion and Diffusion*, Springer-Verlag, New York, 1983.

[34]   Getoor, R. K., and Sharpe, M. J., "Conformal martingales", *Inventiones Math.*, **16** (1972), 271–308.

[35]   Gilbarg, D., and Trudinger, N. S., *Elliptic Partial Differential Equations of Second Order*, 2nd ed., Springer-Verlag, New York, 1983.

[36]   Girsanov, I. V., "On transforming a certain class of stochastic processes by absolutely continuous substitution of measures", *Theory of Probability and Its Applications*, **5** (1960), 285–301.

[37]   Halmos, P. R., *Measure Theory*, Springer-Verlag, New York, 1974.

[38]   Harrison, J. M., and Kreps, D., "Martingales and arbitrage in multiperiod securities markets", *J. Econ. Theory*, **20** (1979), 381–408.

[39]   Harrison, J. M., Landau, H., and Shepp, L. A., "The stationary distribution of reflected Brownian motion in a planar region", *Ann. Prob.*, **13** (1985), 744–757.

[40]   Harrison, J. M., and Pliska, S. R., "Martingales and stochastic integrals in the theory of continuous trading", *Stoch. Proc. Appl.*, **11** (1981), 215–260.

[41]   Harrison, J. M., and Reiman, M. I., "Reflected Brownian motion on an orthant", *Ann. Prob.*, **9** (1981), 302–308.

[42]   Harrison, J. M., and Reiman, M. I., "On the distribution of multidimensional reflected Brownian motion", *SIAM J. Appl. Math.*, **41** (1981), 345–361.

[43]   Harrison, J. M., and Williams, R. J., "Brownian models of open

queueing networks with homogeneous customer populations",
Stochastics, **22** (1987), 77–115.

[44] Hsu, P., "Probabilistic approach to the Neumann problem", *Comm. Pure Appl. Math.,* **38** (1985), 445–472.

[45] Ikeda, N., and Watanabe, S., *Stochastic Differential Equations and Diffusion Processes,* North Holland, Amsterdam, 1981.

[46] Itô, K., *Lectures on Stochastic Processes,* Tata Institute of Fundamental Research, Bombay, 1961.

[47] Itô, K., and McKean, H. P., Jr., *Diffusion Processes and their Sample Paths,* Springer-Verlag, New York, 1974.

[48] Jacod, J., *Calcul Stochastique et Problèmes de Martingales,* Lecture Notes in Mathematics, Vol. 714, Springer-Verlag, Berlin, 1979.

[49] Karatzas, I., and Shreve, S. E., *Brownian Motion and Stochastic Calculus,* Springer-Verlag, New York, 1988.

[50] Karlin, S., and Taylor, H. M., *A First Course in Stochastic Processes,* 2nd ed., Academic Press, New York, 1975.

[51] Karlin, S., and Taylor, H. M., *A Second Course in Stochastic Processes,* Academic Press, New York, 1981.

[52] Letta, G., "Un exemple de processus mesurable adapté non progressif", *Séminaire de Probabilités XXII,* Lecture Notes in Mathematics, Vol. 1321, Springer-Verlag, New York, 1988, 449–453.

[53] Letta, G., *Martingales et intégration stochastique,* Pisa, Scuola Normale Superiore, Quaderni, 1984.

[54] Lévy, P., *Processus Stochastique et Mouvement Brownien,* Gauthier-Villars, Paris, 1948 (2nd ed. 1965).

[55] Lions, P. L., and Sznitman, A. S., "Stochastic differential equations with reflecting boundary conditions", *Comm. Pure Appl. Math.,* **37** (1984), 511–537.

[56] Maruyama, G., "On the transition probability functions of the Markov process," *National Science Report, Ochanomizu University,* **5** (1954), 10–20.

[57] McKean, H. P., Jr., *Stochastic Integrals,* Academic Press, New York, 1969.

[58] Merton, R. C., "On the pricing of contingent claims and the Modigliani-Miller theorem", *J. Financial Economics,* **5** (1977), 241–249.

[59] Métivier, M., *Reelle und Vektorwertige Quasimartingale und die Theorie der Stochastischen Integration,* Springer-Verlag, Berlin, 1977.

[60] Meyer, P. A., "Un cours sur les intégrales stochastiques", in *Séminaire de Probabilités X*, Lecture Notes in Mathematics, Vol. 511, Springer-Verlag, Berlin, 1976.

[61] Papanicolaou, V. G., *The Probabilistic Solution of the Third Boundary Value Problem for the Schrödinger equation and its path integral representation,* Ph.D. dissertation, Dept. of Mathematics, Stanford University, 1988.

[62] Pitman, J. W., "One-dimensional Brownian motion and the three-dimensional Bessel process", *Adv. Appl. Prob.,* **7** (1975), 511–526.

[63] Reiman, M. I., "Open Queueing Networks in Heavy Traffic", *Math. Oper. Res.,* **9** (1984), 441–458.

[64] Revuz, D., and Yor, M., *Continuous Martingales and Brownian Motion,* forthcoming book.

[65] Roberts, A. W., and Varberg, D. E., *Convex Functions,* Academic Press, New York, 1973.

[66] Rogers, L. C. G., "Characterizing all diffusions with the 2M-X property", *Ann. Prob.,* **9** (1981), 561–572.

[67] Rogers, L. C. G., and Pitman, J. W., "Markov functions", *Ann.*

Prob., **9** (1981), 573–582.

[68] Rogers, L. C. G., and Williams, D., *Diffusions, Markov Processes, and Martingales,* John Wiley and Sons, Chichester, 1987.

[69] Royden, H. L., *Real Analysis,* Macmillan, New York, 2nd ed., 1968.

[70] Sharpe, M. J., "Local times and singularities of continuous local martingales", *Séminaire de Probabilités XIV,* Lecture Notes in Mathematics, Vol. 721, Springer-Verlag, New York, 1979, 76–101.

[71] Sharpe, M. J., *General Theory of Markov Processes,* Academic Press, San Diego, 1988.

[72] Stroock, D. W., and Varadhan, S. R. S., "Diffusion Processes with Boundary Conditions", *Comm. Pure Appl. Math.,* **24** (1971), 147–225.

[73] Stroock, D. W., and Varadhan, S. R. S., *Multidimensional Diffusion Processes,* Springer-Verlag, New York, 1979.

[74] Taylor, A. E., *Introduction to Functional Analysis,* John Wiley and Sons, New York, 1958.

[75] Varadhan, S. R. S., and Williams, R. J., "Brownian motion in a wedge with oblique reflection", *Comm. Pure Appl. Math.,* **38** (1984), 405–443.

[76] Wang, A. T., "Generalized Itô's formula and additive functionals of Brownian motion", *Z. Wahr. verw. Geb.,* **41** (1977), 153–159.

[77] Williams, R. J., *Brownian motion in a wedge with oblique reflection at the boundary,* Ph.D. Dissertation, Department of Mathematics, Stanford University, Stanford, California, 1983.

[78] Williams, R. J., "A Feynman-Kac gauge for solvability of the Schrödinger equation", *Adv. Appl. Math.,* **6** (1985), 1–3.

[79] Williams, R. J., "Recurrence classification and invariant measure for reflected Brownian motion in a wedge", *Ann. Prob.,* **13** (1985),

758–778.

[80]   Williams, R. J., "Reflected Brownian motion in a wedge: semi-
        martingale property", *Z. Wahr. verw. Geb.*, **69** (1985), 161–176.

[81]   Williams, R. J., "Local time and excursions of reflected Brownian
        motion in a wedge", *Publ. RIMS Kyoto University*, **23** (1987), 297–
        319.

[82]   Yamada, T., and Watanabe, S., "On the uniqueness of solutions of
        stochastic differential equations", *J. Math. Kyoto Univ.*, **11** (1971),
        155–167.

[83]   Zhao, Z., "Conditional gauge with unbounded potential", *Z. Wahr.
        verw. Geb.*, **65** (1983), 13–18.

[84]   Zhao, Z., "Uniform boundedness of conditional gauge and Schrödinger
        equation", *Comm. Math. Phys.*, **93** (1984), 19–31.

INDEX

Probability and Its Applications

Editors

Professor Thomas M. Liggett
Department of Mathematics
University of California
Los Angeles, CA 90024-15555

Professor Charles Newman
Courant Institute of
Mathematical Sciences
New York University
New York, NY 10012

Professor Loren Pitt
Department of Mathematics
University of Virginia
Charlottesville, VA 22903-3199

Probability and Its Applications includes all aspects of probability theory and stochastic processes, as well as their connections with and applications to other areas such as mathematical statistics and statistical physics. The series will publish research-level monographs and advanced graduate textbooks in all of these areas. It acts as a companion series to Progress in Probability, a context for conference proceedings, seminars, and workshops.

We encourage preparation of manuscripts in some form of TeX for delivery in camera-ready copy, which leads to rapid publication, or in electronic form for interfacing with laser printers or typesetters.

Proposals should be sent directly to the editors, or to: Birkhäuser Boston, 675 Massachusetts Avenue, Cambridge, MA 02139.

Series Titles

K. L. CHUNG/ R. J. WILLIAMS. *Introduction to Stochastic Integration*, 2nd Edition
R. K. GETOOR. *Excessive Measures*
R. CARMONA / J. LACROIX. *Spectral Theory of Random Schrödinger Operators*
G. F. LAWLER. *Intersections of Random Walks*
R. M. BLUMENTHAL. *Excursions of Markov Processes*
S. KWAPIEN / W. A. WOYCZYNSKI. *Random Series and Stochastic Integrals*
N. MADRAS / G. SLADE. *The Self-Avoiding Walk*